# Forests and Urban Environment Planning and Development

# Forests and Urban Environment Planning and Development

R. Dhanai

CYBER TECH PUBLICATIONS
4264/3, Ansari Road, Daryaganj, New Delhi-110002 (India)
Ph.: 011-23244078, 011-43559448 Fax: 011-23280028
E-Mail: cyberpublicationsdelhi@yahoo.com
Website: www.cybertechpublications.com

**FORESTS AND URBAN ENVIRONMENT PLANNING AND DEVELOPMENT**

**R. Dhanai**



First Edition 2011



*Published by :*

**G.S. Rawat for Cyber Tech Publications**
4264/3, Ansari Road, Daryaganj, New Delhi-110002 (India)
Ph.: 011-23244078, 011-43559448 Fax: 011-23280028
E-Mail: cyberpublicationsdelhi@yahoo.com
Website: www.cybertechpublications.com

# CONTENTS

# Preface

The forests covered about half of the earth's surface up to the development of early civilizations but today cover less than one-third of that area. Massive forest clearances were a feature of many early cultures, including ancient Assyria, Babylon, China, Egypt, Greece and Rome. Much of this forest clearance was to provide land for agriculture, although these societies were also voracious consumers of wood for cooking, heating, copper smelting, pottery making, brick-firing, house construction and shipbuilding, and this led to deforestation wherever the forest did not have the opportunity to regenerate. For more than 10000 years the forests of the Mediterranean region have been cleared; they now cover about one-sixth of the region and those that remain, often on land that cannot readily be cultivated, have been degraded by humans and their animals through unmanaged grazing.

The export of high-quality cedar wood from Lebanon to Egypt began nearly 5000 years ago and led, eventually, to the virtually complete destruction of the cedar forests. The clearance of forests is thus not a new phenomenon in human history, except in its present scale and spread. At times, the pressure on the forests eased.

The decline and ultimate fall of the Roman Empire following the sack of Rome by the Goths in the fifth century AD, and the subsequent economic decline of much of Europe, led to the regrowth of forests in some areas. The same happened at the time of the Black Death in the mid-fourteenth century when the population of western Europe was reduced by over one-third, and in some places by half. In Germany, people fled rural life and the oppression of rural overlords into the depopulated cities, while many responded to the invitation of the Polish kings and moved to the eastern territories (the Baltic, East and West Prussia, Pomerania). As a result, the forests of central and western Europe colonized and regenerated over large areas, and it was not until the sixteenth century that population pressure and the demand for forest products again increased to a level comparable to that at the end of the thirteenth century.

*—Author*

# 1

# Global Forest Resources: History and Dynamics

## HISTORICAL PERSPECTIVE

Forests covered about half of the earth's surface up to the development of early civilizations but today cover less than one-third of that area. Massive forest clearances were a feature of many early cultures, including ancient Assyria, Babylon, China, Egypt, Greece and Rome. Much of this forest clearance was to provide land for agriculture, although these societies were also voracious consumers of wood for cooking, heating, copper smelting, pottery making, brick-firing, house construction and shipbuilding, and this led to deforestation wherever the forest did not have the opportunity to regenerate. For more than 10000 years the forests of the Mediterranean region have been cleared; they now cover about one-sixth of the region and those that remain, often on land that cannot readily be cultivated, have been degraded by humans and their animals through unmanaged grazing. The export of high-quality cedar wood from Lebanon to Egypt began nearly 5000 years ago and led, eventually, to the virtually complete destruction of the cedar forests. The clearance of forests is thus not a new phenomenon in human history, except in its present scale and spread.

At times, the pressure on the forests eased. The decline and ultimate fall of the Roman Empire following the sack of Rome by the Goths in the fifth century AD, and the subsequent economic decline of much of Europe, led to the regrowth of forests in some areas. The same happened at the time of the Black Death in the mid-fourteenth century when the population of western Europe was reduced by over one-third, and in some places by half. In Germany, people fled rural life and the oppression of rural overlords into the depopulated cities, while many responded to the invitation of the Polish kings and moved to the eastern territories (the Baltic, East and West Prussia, Pomerania). As a result, the forests of central and western Europe colonized and regenerated over large areas, and it was not until the sixteenth century that population pressure and the demand for forest products again increased to a level comparable to that at the end of the thirteenth century.

Following the Thirty Years War (1618-48), famine and disease again reduced the population of rural Germany, on this occasion by 40%. This pattern was repeated during the Spanish Heritage War (1701-14) and the Seven Years War (1756-63). Following the order of the French King Louis XIV, nearly all towns and castles in south-western Germany were burnt. Their reconstruction consumed an enormous amount of timber so that by the middle of the eighteenth century much of the forest wealth of central Europe had been destroyed, a timber shortage ensued and there was a loss of soil fertility and soil erosion. Vast tracts in northern Germany (e.g. Lüneburger Heide) were degraded to deserts with shifting dunes.

In Britain, where population growth, land clearance and economic development were especially rapid, the natural forests had been heavily depleted by the late Middle Ages. Iron smelting, relying on charcoal, and other wood-based energy users encountered supply difficulties. By the sixteenth century most of the accessible wood resources had been exhausted in England and iron-makers turned to the woodlands of Scotland, Ireland and Wales. Timber began to be imported into England in large quantities for every purpose.

Destruction of the natural forests has thus taken place over most of Europe. It accelerated as the demand for fuelwood and timber grew rapidly in the early years of the Industrial Revolution, but with the continuation of economic growth and industrialization the pressures on the forests have gradually eased. By the early years of the present century, the total area of the European forests had more or less stabilized and in the past decade has even increased slightly. In particular, the pressure on the Mediterranean forests for the supply of fuelwood and land for agriculture and grazing has declined, although the risk of damage from fires continues.

A similar process of forest clearance occurred in the USA (Salwasser *et al.* 1994). At the time of European settlement in the early seventeenth century about 46% of the land was forested; by 1992 this area had declined to about one-third. Huge areas of forest were cleared in the eighteenth and nineteenth centuries as the population grew and industrialization progressed. Early clearing was for agriculture; in 1800, 95% of the people lived on the land, mostly practising subsistence farming. Most wood was used for firewood even in the late nineteenth century, which depleted forest resources especially around population centres. Fencing was the next most important use of wood, followed by early industrial use. The spread of the railroad then further increased demand for wood, which was used for locomotive fuel until the mid-eighteenth century but above all for sleepers or railroad ties. For example, in 1900 it was estimated that the replacement of ties on existing track required the output of 6-8 million ha of forest annually.

The importance of the sea for defence and trade led to laws being passed in several European countries in an effort to protect and expand wood resources for naval vessels and merchant fleets, especially oak for building the hulls and pine for masts. The laws were only partially successful, even for these limited objectives. The forest cover of France had shrunk from perhaps 35% of the country's area at the beginning of the sixteenth century to about 25% by the middle of the seventeenth century, and despite the reforms of the administration of the forests and of the forest laws by Louis XIV and his minister of finance, Colbert, in 1661, the situation continued to deteriorate due to demands for shipbuilding and the smelting of iron ore. Nevertheless these reforms began a forestry tradition and laid the basis for a strong forest service.

A similar realization that the forest resource was not limitless led to the emergence of rules to control the use of forests in Germany, Denmark and The Netherlands in the eighteenth century, in India in the mid-nineteenth century, in Sweden and Finland in the late nineteenth century and in the USA in the early twentieth century. At that time colonial powers were reserving forests, introducing forest laws and establishing forest services in East Africa, Malaysia and many other now-independent countries. In the eighteenth century, efforts to restore forest cover and guarantee future timber supplies were made. Today Germany has 30% forest cover, which originated in the plantations of the seventeenth century; initially these stands were largely of single species (oak, beech, Scots pine and Norway spruce) but today are slowly being converted into mixed stands with uneven age structure.

The development of management techniques for natural forest began in Europe for a number of purposes. Sometimes this process was accelerated for specific objectives, for example to reduce the risk of avalanches, to control erosion, to stabilize sand dunes or to protect agricultural crops, as well as to provide firewood for towns. Management planning in the sense of regulation of yield had been introduced in France by Colbert, while in Denmark management plans with rotations of oak of 100 years were instituted in 1763.

Originally recruited from retired soldiers, the foresters of the specialized forest services created by the rulers of various German states (e.g. Brandenburg-Preussen, Hanover, Saxony) were to have a profound influence on forest science and practice even up to the present. The experience they gained, combined with traditional forestry knowledge, was published and widely disseminated. One of the best-known publications of that time was *Sylviculture! Oeconomica*, published in 1713 by Hans von Carlowitz (1645-1714). Von Carlowitz was a mining engineer in Saxony preoccupied with the insufficient timber supply for the Saxon mines. His book dealt not only with timber harvesting, species selection, utilization of timber and sowing of conifer seeds to reforest clear-fellings, but also with promoting the introduction of foreign (exotic) species and he was the one who first coined the term *Nachhaltigkeit* (sustainable forest management). The handful of German foresters who became the founders of forest science included Heinrich Cotta (1763-1844), Wilhelm Pfeil (1783-1859), Johann Christian

Hundshagen (1783-1834), Carl Heyer (1797-1856), Gottlob König (1779-1849) and George Ludwig Hartig (1764-1837), who as professor at the University of Berlin introduced forestry training into the Prussian forest service and had a strong impact on the forest management principles of his day, not only in Prussia but all over central Europe. Hartig's rule was that no more nor less timber should be extracted annually from state forests than could guarantee perpetual supply, which became the principle of central European forestry, i.e. sustainable forest management.

The reduction in forest area had been halted or even reversed in the industrialized countries by the 1920s. Since then the standing volume of forests has increased with better management and increased protection, and in Europe the yearly forest harvest of many countries is smaller than the annual increment. The reasons for this historical transition from rapidly diminishing to stablenor even increasing, forest areas in many countries of the Northern Hemisphere with industrial or service economies are complex. The substitution of coal for fuelwood for industrial and domestic use was one obvious factor. However, more important were the major demographic and economic changes that took place as industrialization and economic growth proceeded.

Rural populations fell drastically as people migrated to the cities and towns. By the turn of the present century, the average proportion of the population living in the rural areas of the industrial countries was around 40% compared with 80% or more at the beginning of the Industrial Revolution. Today, in Europe and the USA, the rural population is about 25% of the total. National population growth rates have dropped to around replacement level in most of the industrial countries. Equally important were the changes that took place in agricultural production. Subsistence farming gave way to an increasingly capital-intensive form of agriculture relying on mechanization and high inputs of fertilizers and pesticides and, in animal husbandry, the stall-feeding of animals. Rather than requiring ever-expanding amounts of land, agriculture became more productive, leading eventually to the present politically charged problems of crop surpluses and the need to 'set aside' or take increasing amounts of farmland out of food production.

Additionally, increased public awareness of the value of forests and of environmental issues in urbanized societies of the postindustrial or service economies, combined with increasing demand for new services from forests such as recreation and amenity, have been important recent trends in stabilizing forest areas.

Although history does not repeat itself exactly, the experience of the industrial world helps shed light on what is happening in many developing countries in relation to the pressures on their forests. Expanding populations (especially those existing at subsistence level), accompanied by significant rural-urban migration, result in increased demand for agricultural production. Even highly punitive forest protection legislation has been largely ineffectual in the face of the desperation of people trying to grow enough food, or graze sufficient animals, to survive.

It is only with increased agricultural production supplying stabilized populations that the pressure to clear forest land diminishes. Indeed, sustainable forestry may be possible only where agriculture is, if not sustainable, at least stable. This transition can be seen taking place today in some of the rapidly industrializing countries of the developing world. The Republic of Korea, for example, where the rural population has fallen sharply while agricultural production has increased dramatically, has reached the turning point at which the area of forest can begin to expand again, aided by massive reforestation programmes. In others, such as the newly industrializing countries of the Pacific Rim, forest depletion continues, although the economic and demographic conditions and social demands for stabilizing the remaining areas of forest are beginning to emerge.

In many other developing countries, however, rural population growth continues virtually unabated, agricultural productivity is low and economic stagnation means that there are few employment opportunities for those wishing to leave the land. In such cases, increasing pressure to clear forest for agriculture is inevitable whether legalized by land reform or not.

The parallels with the experience of the industrial world are thus obvious but what makes today's position in the developing world different is the sheer impact of population numbers, the pace of change and the fragile nature of many tropical soils and their inability to be used for long-term agriculture. Regarding population numbers, for example, the population of the European continent grew from about 140 million in 1750 to 265 million by 1850, 392 million by 1950 and 499 million by 1990. The population of the USA grew from 23 million in 1850 to 151 million in 1950 and 248 million in 1990. Today, on the other hand, around 50 million people are being added to the rural populations of the developing world *every year.* Where there were about 12 ha of forest per head of the global population of about 500 million people in 1750 there is now about 0.75 ha per head, and the rate at which the forests are disappearing, especially tropical forests, is far faster than at any other time in human history, despite net gains of forest cover in some developed countries and some evidence of a slightly slower rate of deforestation in developing countries in recent years.

## RECENT ESTIMATES OF GLOBAL FOREST AREA

The following section is based on 1990 baseline figures, updated to 1995, prepared by the Food and Agriculture Organization (FAO) Forest Resource Assessment Programme, reported in and largely drawn from *State of the World's Forests 1997* (FAO 1997a).

In 1995 forests were estimated to cover 3454 million ha, or 26.6% of the total land area of the world (Greenland and Antarctic excepted). Almost two-thirds of the world's forests were located in seven countries: Russia, Brazil, Canada, USA, China, Indonesia and Zaire; 29 countries had more than half of their land covered by forest, of which 21 were in the tropical belt. However 49 countries, in addition to the many non-forested small island states and territories, had less than 10% of their land covered by forests. Five entire subregions were in this category: North Africa (1.2% of the land area), Near East (1.9%), Temperate Oceania (6.2%), Non-tropical Southern Africa (6.8%) and West Sahelian Africa (7.5%).

The latest information on the distribution of forest and of forest cover change by ecological zones was provided in 1990; the results of the Forest Resources Assessment 2000 were released by FAO in 2000. In 1990, temperate and boreal forests occupied 1.64

billion ha and tropical forests 1.76 billion ha. A breakdown by ecological zone was available only for tropical forests.The figures on tropical forests show that most (88%) are in lowlands; of these, tropical rain forests accounted for 47% of all lowland tropical forests, followed by moist deciduous forest (38%) and dry and very dry formations (15%). The loss of forest cover through deforestation is discussed later.

## Natural forests

The forest area estimates described above include undisturbed forests, forests modified by humans through use and management (or 'seminatural' forests) and forests created artificially by humankind (i.e. forest plantations) by afforestation or reforestation. (Afforestation is defined as the establishment of a tree crop on an area from which it has always, or for a very long time, been absent. Reforestation is defined as the establishment of a tree crop on forest land.) In most industrialized countries, particularly in continental Europe, forests are being managed in such a way that at management-unit level a continuum exists from low-intensity management, involving natural regeneration, through more intensive methods involving some artificial planting to highly intensive methods with complete planting and cultivation; this makes it difficult to isolate figures for natural forest and plantations. The distinction between natural or seminatural forests and forest plantations can more easily be made for developing countries and some industrialized countries such as New Zealand in which forest plantations have been established using introduced species.

Interest in natural forests, particularly their role in the conservation of biological diversity, has led to efforts to compare forests today with what is thought to be their original character and to give complete protection to areas of forests which have had no, or minimal, human interference. Although there are difficulties in identifying the extent of natural forest, compounded by problems of definition, some information exists that can be used as an indication of broad patterns of natural forests in various regions.

An attempt has been made by the World Wide Fund for Nature (WWF) to quantify the area of forests in western Europe that has been relatively undisturbed by humans or which has retained much of its natural character. Their report (WWF 1994) distinguishes between 'virgin forest', defined as 'forest ecosystems whose characteristics are determined exclusively by natural location and environmental factors... without human influences present or visible any more', and 'natural and ancient seminatural forests', which 'have not been planted or sown by man for the past two centuries' and 'which continue to have a large number of the natural elements'. The study found that only a small proportion (probably < 1%) of the total forest land in northern and western Europe could be considered as virgin forest, which has arisen since the last glaciation. Almost all was located in Sweden, Finland and Norway, with small areas in Greece, Austria and Switzerland and (according to another author) in France. In eastern Europe, Slovakia and Belarus have considerable areas of virgin forest while Poland and Croatia have small areas. In addition, in northern and western Europe, the WWF report identifies natural and ancient seminatural forests representing 2.1% of the total forest cover (1990) of the 16 countries concerned. There are a further 3 million ha of land in the region in national parks and other protected areas and another 50 000 ha in small forest reserves (mostly for nature conservation and scientific research), whose use is tightly restricted.

The situation in temperate and boreal North America is quite different from that of densely populated Europe and Japan, where use and management of forests for many centuries have left very little of the original forest area untouched. 'Old growth forests', as they are called in North America, still cover extensive areas. On the lands managed by the US National Park Service alone, old-growth forests covered 1.97 million ha in 1988. Although not a direct measure of the extent of old growth forest, it may be noted that the area of forest included in national parks and other protected areas in North America (USA and Canada) was reported to be nearly 49 million ha in 1990.

The FAO Forest Resource Assessment 1990 did not distinguish between undisturbed and disturbed 'natural' forests in developing countries. However, the Forest Resource Assessment 1980 made estimates of the areas of undisturbed closed forests (primary forests and old secondary forests where there had been no logging for the last 60-80 years) and of closed forests that were included in national parks and other protected areas (thus relatively undisturbed, at least in theory). At that time, these two categories together represented 60% of the total closed forest area in the tropics, a proportion varying from 39% in tropical Asia to 59% in tropical Africa and 69% in tropical America. These different proportions by region reflected a slower development of large-scale harvesting in tropical America compared with tropical Africa and Asia, and also the fact that, in tropical America, spontaneous colonization did not follow in the wake of logging as systematically as in the two other regions due to lower population pressure. Although the two categories do not match the concept of 'virgin forests' of Europe and of 'old growth forests' of North America discussed above, the sum of the two categories nevertheless gives an indication of the amount of forest disturbance (or management, depending on the point of view of the observer) that existed around 1980 in the humid tropics. Although corresponding estimates for 1990 and 1995 are not available, it is likely that the share of undisturbed forests remains higher in the three tropical regions than in Europe and probably in North America too.

## Forest plantations

Recent data in this section are also available in *State of the World's Forests 1999* (FAO 1999), which also discusses some of the issues concerned with forest plantations in more depth.

### *Early development*

By the seventeenth century the decline in the area of native forest in European countries led to the planting of trees, largely to provide alternative sources of timber supply. Sometimes plantations were established for the provision of services or other products than timber. They might be planted as shelterbelts, for dune stabilization, for amenity or for the supply of firewood.

Plantations of native species were established at first in areas where forests occurred naturally, for example France, Germany, England and Scotland, but later also on previously unforested land. In France, for example, planting of *Pinus pinaster* was started on the sand dunes of the Landes and with *Pinus sylvestris* and *Picea abies* on former agricultural land in the Vosges; in Germany likewise agricultural land was planted with Norway spruce *(Picea abies)* in Saxony. In tropical countries such as Myanmar (then Burma), teak *(Tectona grandis)* was planted as a native species in the *taungya* system.

Teak was one of the first exotic forest plantation species to be used, being planted in Sri Lanka and the island of Java (Indonesia) from early in the nineteenth century. Exotic species, such as Douglas fir *(Pseudotsuga menziesii)*, Sitka spruce *(Picea sitchensis)*, Japanese larch *(Larix kaempferi)*, lodgepole pine *(Pinus contorta)* and poplar *(Populus deltoides)*, were introduced to Europe during the nineteenth century, when they began to play an increasingly important role in forest plantation programmes. Other exotic species, particularly the eucalypts *(Eucalyptus* spp.) and wattle *(Acacia* spp.) from Australia, were introduced as exotics in tropical and subtropical countries from the middle of the nineteenth century, while *Pinus radiata* was introduced from California to New Zealand and other countries such as Chile from the early years of the twentieth century. Box 1.3 describes in detail the experience with forest plantations in Denmark.

The main reason for the establishment of plantations remained the decline in natural forest area and a scarcity of wood and, as discussed above, the conversion of woods and forests to agricultural and grazing use. In recent years increased areas of natural forest have been managed for nature conservation, recreation, wildlife parks, etc., and commercial wood production has been either reduced or eliminated. For example, over 25 000 ha of high-yielding hybrid poplar plantations have been established in the north-western USA between 1992 and 1997 in response to both increased demand for poplar wood for orientated-strand board and decreased supply from public forests (Anon. 1996). On the other hand, some new areas have become available in European countries as land is taken out of agriculture due to trade and market considerations.

### *Objectives*

Forest plantations are tree crops that are in some, but not all, ways analogous to agricultural crops. They often have a simple structure, at least in youth, and are usually composed of one or a few species (but not varieties as in agriculture, except in a few cases such as the intensively bred poplars) chosen for their fast growth, yield of specified products and ease of management. When established for wood production they have a higher productivity of usable wood than natural forests, but due to the way they are managed they do not, indeed cannot, provide the full range of goods and services that natural, seminatural or even secondary forest can provide.

Although the objective of many plantations is the production of industrial roundwood and/or fuelwood, many are established for environmental protection or other services (e.g. soil and water conservation, enhancement of agricultural production in agroforestry systems, carbon sequestration) and some also provide non-wood forest products, such as fodder, various foodstuffs, medicines, etc.

*Area*

The area of forest plantations throughout the world started to increase in the 1970s as many governments became concerned about wood supplies for industry, and in developing countries fuelwood supplies, and has continued to increase since. However, there are no reliable global figures for plantation areas because forests of native species in several developed countries in the temperate and boreal regions, especially in continental Europe, are frequently regenerated naturally and it is not possible to distinguish those areas where supplementary artificial planting has been done. Furthermore many countries consider their plantations as 'seminatural' forests over a certain age, because as stands mature the clear initial row layout of the trees is lost and other species naturally regenerate under the canopy. For example, Austria, Czech Republic and Finland, in responding to a questionnaire to collect data for the ECE/FAO Temperate and Boreal Forest Resource Assessment (TBFRA), stated that they had no plantations in their countries as defined by the TBFRA process.

In 1995, an approximate estimate of the area of plantations in developed countries was 60 million ha, comprising 13.7 million ha in North America, 22.2 million ha in the Commonwealth of Independent States, 12.1 million ha in Europe and 13.2 million ha in Oceania (Australia, New Zealand, Japan). The most significant areas of plantations were in the Russian Federation (17.3 million ha, 2.1% of the country's total forest area), the USA (13.7 million ha, 6.3%), Japan (10.7 million ha, 44.4%), Ukraine (4.4 million ha, 46.8%) (FAO/ECE 1996), Spain (1.9 million ha, 14.9%), New Zealand (1.54 million ha, 19% of the total forest area in 1996, 91% of which was *Pinus radiata)* (Anon. 1997a) and Australia (1.04 million ha in 1994, 85% of which was softwood species, mainly *P. radiata)* (Anon. 1997b).

The estimated 'net' plantation area of 55 million ha in developing countries in 1995 was about 2.8% of the total area of forests in developing countries. In 1980 the net plantation area was assessed at about 40 million ha. It has thus increased by about 15 million ha in 15 years, but even this figure may be liable to error. Many developing countries, especially those with large forest plantation programmes, provided updated information on their present plantation plans to FAO in 1996 and 1997 from which the reported annual rate of new plantations of 3 million ha was derived. Note, however, that this is the reported or planned rate, which may not necessarily have been achieved. Most of the countries with large plantation estates indicated that they intended to double their plantation areas between 1995 and 2010.

It was estimated from reported figures that 57% of the forest plantation area consisted of hardwood species and 63% was established for industrial purposes. Nearly three-quarters of these plantations were in the Asia-Pacific region, where China (21 million ha) and India (20 million ha) dominate, while about 15% were in Latin America and 10% in Africa.

One of the trends in the tropics has been that the proportion of industrial plantations established in large blocks fell from 40% of the total plantation area in 1980 to 35% in 1990. The proportion of smaller plantations established through farm forestry or agroforestry programmes grew in importance during the period 1980-95, particularly in the Asia-Oceania region. Unfortunately, the figures on small-scale private or community-owned forest plantations are even less reliable than for large-scale plantations. Some of these farm forestry or agroforestry plantations supply industrial wood markets for pulpwood (e.g. Brazil and Thailand) or peeler logs (e.g. India). In many countries, particularly those with limited forest area, planted trees grown outside the formal forest area often provide the bulk of fuelwood, poles, construction wood, utility wood, as well as fodder and other non-wood forest products for household use.

Of the area of hardwood plantations planted for industrial use, 30% or nearly 10 million ha consists of eucalyptus, followed by acacias (3.9 million ha or about 12% of the hardwood area) and teak (about 7%). Short rotation plantations of hardwood species such as the eucalypts, acacias and *Gmelina arborea* have been grown for many years by the private sector, but the establishment of plantations of teak or other valuable hardwood species has been carried out only by government forest services because of their slow growth and hence delayed returns. However, the likelihood of reduced supplies of high-quality hardwood logs derived from natural forests, combined with increasing purchasing power and expected higher prices for logs, is leading to increasing interest in investment by the private sector in valuable hardwood species, especially teak, in a number of countries, for example India, Malaysia, Costa Rica and Ghana.

Of the softwoods grown for industrial purposes, fast-growing pines such as *Pinus radiata, P. patala and P. caribaea* constitute about 25% of the area while other, often slow-growing pines (e.g. *P. kesiya, P. massoniana, P. merkusii, P. rox-burghii, P. halepensis, P. pinaster* and *P. wallichi-ana)* make up about 36%.

Responsibility for the monitoring and regulation of plantation crops grown for food and certain other purposes has long been the job of the agricultural sector. In recent years, however, several of these crops (the main ones being rubber, coconut and oil palm) have been providing 'forest' products that have been used for wood and fibre. For example, rubber wood is now used for the manufacture of about 80% of the furniture made in Malaysia, while coconut and oil palm trunks and the branches of rubber wood are used for various forms of reconstituted 'wood'. Rubber wood and coconut stems are derived from the conversion of old plantations formerly disposed of by burning, while oil palm fruit residues are used for medium-density fibre board. The development of these new markets has thus not only improved financial returns but also used the resources in an environmentally friendly manner.

The area of these species appears to have increased from the 14 million ha reported in 1990. The increase may be due to better coverage of the data, although it is known that oil palm areas are increasing rapidly, rubber tree areas are also increasing and coconut plantations are decreasing. Coconut plantations comprise the largest area (about 42% of the total), rubber 36% and oil palm 22%. Most of the coconut plantations are in Indonesia (33% of the area) and the Philippines (28%); most of the rubber plantations are in Indonesia (34%), Thailand (20%) and Malaysia (18%), while most of the oil palm plantations are in Malaysia (44%) and India (29%) (see Table 1.3). Not all of the areas mentioned above are suitable or available for substitute 'timber' production but they illustrate the potential.

### *Contributions of forest plantations to wood supply*

The continuing increase in the area of forest plantations which have been established for industrial wood supply has been to meet the reduction of outturn foreseen from natural forests arising from deforestation and changes in land use (largely in the tropics and subtropics) or from natural forest being taken out of production and devoted to service functions such as conservation. It has been believed that the outputs from forest plantations can help to reduce the pressure on natural forests as sources of industrial wood supply; while the logging of tropical natural forests is not the prime cause of deforestation, logging roads often provide the means for farmers to gain access to forests. The reduction of logging, combined with effective protection, may thus help to reduce deforestation in certain locations until land use and ownership are clarified. However, none of these palliatives will remove the underlying causes of deforestation: high rates of population growth, poverty, hunger and a shortage of fertile land to cultivate.

The potential of forest plantations to meet demand for industrial roundwood is considerable; it has been estimated that the present global demand for paper pulp could be met from an area equivalent to only 1.5% of the world's closed forest area (IIED

1996). No global estimates of current output of timber from forest plantations are available, although FAO's global fibre supply model (GFSM) (FAO 1998) estimated that the potential annual growth of industrial wood from forest plantations in developing countries was about 5% of the increment of natural forests in 1995. In some countries, plantation production already makes a highly significant contribution to the industrial wood supply, for example in New Zealand 99% of industrial roundwood in 1997 was grown in plantations, while in Chile the equivalent figure was 95%, in Brazil and Argentina 60%, and in Zambia and Zimbabwe 50%.

Estimating the future contribution of forest plantations to wood supply is at present imprecise and is based on many more or less unreliable assumptions, particularly concerning the rate at which afforestation will continue. By the year

2010 the GFSM (op. cit.) estimated that the potential increment from forest plantations would be about 40% of that from natural forests in Asia, Oceania and Latin America and about 15% in Africa, under rates of deforestation and afforestation largely the same as today.

## CHANGES IN FOREST COVER AND CONDITION

An analysis of changes in the cover and condition of the world's forests requires a differentiation between:

1 the increase in forest cover (by afforestation or natural colonization of trees on non-forest land) or decrease (by deforestation) of forest area; and

2 changes in forest condition, either positive (recovery of degraded stands, stand improvement treatments) or negative (decline, defoliation or dieback, effects of forest fires, degradation through unsustainable exploitation for wood, overgrazing, effects of pests and diseases).

### Changes in forest cover

Between 1980 and 1995, the extent of the world's forests decreased by some 180 million ha, an area about the size of Indonesia or Mexico. This represents a global annual loss of 12 million ha, an area equivalent to the size of Greece or Bangladesh. During this 15-year period, developing countries lost nearly 200 million ha of natural forests, mostly through clearing for agriculture (shifting cultivation, other forms of subsistence agriculture, the establishment of cash crop plantations such as oil palm, and ranching). This was only very partially compensated for by the establishment of new forest plantations. Over the same period, forests in the developed world expanded slowly (by some 20 million ha) through afforestation and reforestation, including natural regrowth on land abandoned by agriculture.

Forest plantation and natural regeneration on abandoned agricultural land more than compensated for clearing of forests due to urbanization and infrastructure development in most industrialized countries, outside the former USSR. While the gain of forest cover in developed countries was 20 million ha in the period 1980-95, 9 million ha of this increase occurred in the 5 years from 1990 to 1995.

The situation is quite different in the developing world, where deforestation exceeded net afforestation/reforestation, particularly in the tropical zone. As a whole, the annual rate of deforestation in the developing world between 1990 and 1995 was 0.7%, equivalent to 12.6 million ha, with the highest rate in tropical Asia-Oceania, closely correlated with population and income growth. The lowland forest formations were the most affected by deforestation, although the proportional loss during this period was greater in upland formations (1.1% compared with 0.8% in the lowland formations). There is some evidence that the rate of loss of forest cover in developing countries was slowing towards the end of the 15-year period 1980-95. The annual rate of forest loss in the period 1980-90 was 15.5 million ha compared with 13.7 million ha between 1990 and 1995.

## Conversion of forests to other land cover

The Forest Resource Assessment 1990 carried out an assessment of the relative importance of the various factors involved in deforestation at regional and global levels between 1980 and 1990 in the whole tropical belt. Among the most significant outputs of the study were the 'area transition matrices' of the type, which indicate transfers from one land cover class to another over the 3068 million ha of the tropical zone covered by the sample. For instance, the first row shows that 1275.9 million ha of the total area of 1368 million ha of closed forest in 1980 remained as closed forest up to 1990 and also shows the fate of the 92.1 million ha converted to other land cover classes (open forest; long fallow; fragmented forest; shrubs and short fallow; other, i.e. non-wooded, land cover; and forestry or woody forest and agricultural tree plantations). All these transfers involved changes of woody biomass attempts to capture by replacing each class along the *y*-axis (average biomass per hectare) and showing the importance of transfers along the *x*-axis.

## Changes in forest condition

Factors affecting the health and vitality of all types of forests have attracted increasing attention in recent years. Outbreaks of fires have made headline news in developed and developing countries alike, while insect and disease attack and generalized declines in forest condition have caused more localized concern. Insects and disease and forest fires exist under 'natural' systems and conditions; what has caused concern has been their spread in artificial systems and in conditions considerably modified by humankind, resulting in direct economic loss.

Fire arising from natural causes (e.g. lightning) has been a major influence and is a driving evolutionary force in several forest ecosystems, such as the pine forests of Central America. Fires are important to the health and maintenance of these ecosystems and the total exclusion of fire from them can lead to the build-up of fuel and ultimately abnormally destructive fires that cause the loss of the ecosystem, as happened with the fire exclusion policy of the National Parks Service of the USA and the disastrous fires of 1988. However, many fires arising from human activities, whether planned or unplanned, have damaged and destroyed large areas of forests and woodlands. Resource managers face a demanding public often with conflicting needs and incomplete information, leading to the development of new policies that take into account public attitudes towards fire management.

At the global level, an agreement was signed at a meeting in 1990 to initiate the International Decade for Natural Disaster Reduction and the target date of 2000 was set for signatory countries to effectively monitor and manage wildfires in their respective countries. In view of the lack of information on causes, extent and the number of fires at present, and the difficulties in collecting data, this target date is unlikely to be met.

*Developed countries*

Total forest area is slowly increasing in the developed world, although the need for protection from fire, insects and disease has continued and some aspects of forest condition give cause for concern. Although the widespread decline and death of certain European forests due to air pollution predicted by many in the 1980s did not occur, deteriorating forest condition has remained a serious concern in Europe and North America. The main causes of the decline of forest condition in Europe have been low soil moisture availability due to drought, and high temperatures. These factors, combined with airborne pollution and the use of non-adapted seed sources in forest plantation development, have predisposed forests to attack by insects and disease and to decline. Forest damage due to air pollution is severe in some parts of central and eastern Europe in particular. Air pollution is also said to have a role in the decline observed in certain species in other industrialized countries, such as sugar maple *(Acer saccharum)* in eastern Canada, the high-elevation forests of red spruce *(Picea)* in eastern USA and *Cryptomeria japonica* stands in Japan.

The exclusion of fire may have caused the decline of some forest ecosystems, such as oak stands in central USA or natural eucalypt forests in Australia where fire is an integral part of the ecology of some plant communities and here it may be used to facilitate their natural regeneration.

Despite an overall increase in the number of fires (by some 40% in Europe, 8% in North America and 120% in the former USSR), wildfires over large areas of forest have become less frequent in the industrialized countries. Further, the average area burned by wildfire decreased in most regions between 1983-90 and 1991-94. There has been a slight reduction in total areas burned in the developed world overall as a result of improved prevention, detection and control systems, although there has been an increase of 15% in the former USSR. In 1990, the average area of forest and other wooded land affected annually by fires in Europe, North America and the former USSR together was about 4.26 million ha or 0.22% of their total area of forest and other wooded land. Pests and diseases remain constant threats, particularly to the semi-natural forests and plantations in the industrialized world. Trees are more likely to be attacked when under stress, whether from abnormal weather conditions such as drought or high temperatures, airborne pollution, uncontrolled movement of germplasm or lack of management. Recent outbreaks of pests and diseases that may have been related to stress include the decline of various species of oaks in many parts of Europe and in central Russia, the recurrent infestation of forests in Poland by the nun moth, the attack on beech by the beech scale in western and central Europe, birch and ash diebacks in north-eastern USA, the littleleaf disease of shortleaf pine in southern USA, and the 'x-disease' of pines in southern California.

*Developing countries*

Less attention has been paid to forest condition in developing countries because reduction in forest area has attracted more attention. This reduction in density has often been the result of over-exploitation for timber and fuelwood. Overgrazing and repeated bush fires are other significant causes of decrease in density, especially in the dry tropical and non-tropical zones.

Every year, very large areas of savannah woodland and mixed forest-grassland formations are affected by fires set by herders to provide an early flush of green grass when the rains start, particularly in the dry zones of Africa and South America, although no reliable data are available on their extent.

Forests in the humid tropics have also, at times, been affected by large fires, the most serious in recent years having been those associated with the dry weather conditions arising from the El Niño Southern Oscillation (ENSO). In 1997-98, these caused large-scale destruction of logged and secondary forest in Indonesia (particularly Kalimantan, Sumatra and Irian Jaya), Mexico and Central American countries, as well as leading to smoke pollution over wide areas beyond the borders of these countries. The cause of these fires has been burning for land clearance, but it should be noted that the fires are not necessarily in forest; although described as 'forest fires' they are frequently in grassland or on land that has been cleared of forest. It is *estimated* that of the 2 million ha burned in Indonesia in 1997, 150000-200 000 ha may have been in forest. Other serious fires associated with ENSO effects have included that in East Kalimantan, which burned 3.6 million ha in 1983.

Coniferous forests in the humid tropics have often been affected by fires, although it must be noted that fire is frequently necessary to maintain and regenerate them. In the 1980s, the area of pine forest in Honduras and Nicaragua burned annually amounted to some 65 000ha (or about 3.5% of the total pine forest area of these two countries), and widespread fires in natural and artificial tropical pine forests occur in other countries such as Guatemala, Mexico and Indonesia (northern Sumatra).

In the subtemperate and temperate zones of the developing world, fire is also a permanent threat to forests, particularly when they are no longer used by local people for grazing and other purposes. In the 1980s, the average area of forest and other wooded land burned annually was 140 000 ha in the temperate/subtemperate zones of South America (including southern Brazil). From 1950 to 1990, fires in China are reported to have affected an average of 890 000 ha annually, the most damaging one having been the 'May 6' fire, which burned some 1.85 million ha in the north-eastern province of Heilongjiang in 1987.

In the absence of a global statistical fire database, it is difficult to provide an overall estimate of the annual extent of fires in forests and other wooded lands.

A very crude estimate for the temperate/subtemperate and humid tropical zones of the developing world (leaving aside the significant dry tropical zone, for which little reliable information exists) would be of the order of 2 million ha of forest and other wooded land annually during the 1980s. Given the lack of sufficient capacity in fire prevention and control in most developing countries, no significant reduction in wooded areas burned is likely to occur in the near future.

Outbreaks of pests and diseases in developing countries are generally reported for those plantations and planted trees where the impact is most apparent. Introduced pests are usually extremely destructive and several of them have had very damaging effects in recent years in the developing world. Examples of introduced insects affecting plantations and planted trees include the *Leucaena* psyllid, *Heteropsylla cubana,* which has spread into Asia and the Pacific islands and is now extending across Africa; the cypress aphid, *Cinara cupressi,* which is established in eight eastern and southern African countries and is causing heavy mortality among a number of exotic and indigenous species but especially the important plantation species *Cupressus lusitanica,* threatening its future role in the plantation programmes of these countries; and the European woodwasp, *Sirex noctilio,* which has spread into Argentina, Uruguay and southern Brazil, affecting principally *Pinus taeda* but which may become a threat for the large Chilean plantation estate of *Pinus radiata.*

Strategies to control introduced insects from attacking forest trees include regulatory measures, eradication, integrated pest management, and monitoring of population levels and occurrence of potentially harmful species.

Effective monitoring of insect populations will be especially important in view of the large plantation programmes being carried out to produce industrial roundwood, which often rely on one or a few exotic species. Such programmes may be able to afford the investment that is necessary for control measures. Smaller-scale programmes may have to insure against failure due to unexpected attack by insects or diseases by diversifying into several species adapted to the sites and end-uses.

There is also evidence of forest decline due to a combination of biotic and abiotic factors in the developing world, with air pollution likely to play an increasing role in some cases as industrial and transportation infrastructure develops.

Examples of such decline include neem *(Azadirachta indica)* in the Sahel, framiré *(Terminalia ivoren-sis)* in Ivory Coast and Ghana, and *Eucalyptus globulus* plantations in Colombia and Peru, while airborne pollution is affecting forests near large cities and industrial areas in China (e.g. *Pinus massoniana* stands near Nanshan). Despite the overall significance of the corresponding damage and losses, surveys of forest decline and diebacks in developing countries remain all too rare.

## CONCLUSIONS

The world's forests cover an area of 3454 million ha or approximately 26.6% of the land surface, with 56.8% of this area in developing countries, which are mostly tropical. Forest cover has largely stabilized in most industrialized countries but deforestation continues in many developing countries. Between 1990 and 1995, the area of natural forests in developing countries decreased by an estimated 13.7 million ha per year, although this rate of loss appears to be slightly less than in the period 1980-90. Furthermore, forest management is relatively little practised at present in natural tropical forests. The extent and condition of the global forest resource are determined by many economic, social and political factors external to the forestry sector, including, in particular, continued population growth and higher rates of global economic growth. Population growth, which will occur mainly in tropical developing countries, will continue to be combined with urbanization, while global economic growth will continue to be combined with changing consumption patterns, especially in the regions with fastest economic growth.

In the coming decades, pressures for increased food production are expected to lead to continued conversion of forest land to agriculture in many developing countries. It is estimated that in developing countries 90 million ha of land, of which about half may be forest land, would need to be converted to arable crop production alone between 1990 and 2010 (FAO 1995). Agricultural land expansion is projected to be faster in sub-Saharan Africa than in the past; given the unsuit-ability of much of this zone for agriculture, there must be a continuing threat to natural forest cover from agricultural expansion as land continually goes out of agricultural production. Infrastructure development will also contribute significantly to the continued loss of forests. In addition to deforestation in developing countries, large areas of forest worldwide are being degraded by overharvesting, overgrazing, pests, disease, wildfires and airborne pollution.

At the same time as forest cover globally is decreasing and forests are being degraded, demands on forests to supply wood and non-wood products and social and environmental services are increasing. Global consumption of wood increased by 36%

between 1970 and 1994, and is expected to increase by another 20% by 2010. Greater emphasis is being put on the services that forests and trees can provide, including soil and water conservation, sequestration of carbon for the mitigation of climate change, conservation of biological diversity, support in combating desertification, enhancement of agricultural production systems, improvement of living conditions in urban and peri-urban areas, and provision of educational and recreational opportunities. Forests will remain an essential source of the livelihood of the poorer sectors of the world's population and a home to indigenous peoples for some time to come. The challenge of meeting the growing demand for forest products while safeguarding the ability of forests to provide a wide range of environmental services will increasingly be met through the planting of trees, either within the forest or outside. However, less natural or seminatural forest will be converted to plantations due to the emerging values of such ecosystems, and suitable land for plantation development will therefore be in short supply in many places. Governments will thus aim to encourage large- and small-scale landowners to plant trees outside forests, often integrated into agricultural systems, through policy measures including incentives. There are likely to be moves towards making blocks of trees more 'natural' with a diversity of species, in some instances in order to provide a wider range of goods and thus serve as an insurance against the possibility of a single species failure or to guard against possible loss of soil fertility or site degradation, as well as to provide for improved amenity and recreational potential. There should be less emphasis on the area or quantity of forest cover and more attention paid to forest health and condition.

Forest cover in industrialized countries will continue to expand, and several newly industrialized countries will see their forest cover stabilize and even increase. Industrialized and newly industrialized countries will place greater emphasis on the conservation of natural forest where it still exists and on the conservation of seminatural forest. Plantation areas will continue to expand, either as intensively managed systems akin to farming practices or management will tend to move towards a more 'ecosystem' approach.

# 2

# Forest Types and Classification

## INTRODUCTION

Closed canopy forests and savannahs occupy a tremendous amount of land, approximately 3540 million ha (Perry 1994) or about 26% of the earth's terrestrial surface (FAO 1997). Tropical forests comprise approximately 35% of the total, temperate forest 17%, boreal forest 16%, and the remainder is savannah and open woodlands. Among countries and continents, about 520 million ha of forest are in Africa (15.1%), 465 million ha in Asia and Oceania (16.4%), 146 million ha in Europe (4.2%), 950 million ha in South America, Central America and the Caribbean (27.5%), 457 million ha in North America (13.2%) and 816 million ha in the countries of the former USSR (23.6%). There is tremendous floristic diversity in these forests, with more than 10000 known species of trees and shrubs, and tens of thousands of species of non-woody vascular and non-vascular understorey plants and epiphytes. The tropical forests are the most diverse, with over 2500 species of trees in both the Malay Peninsula and South America for example.

The tremendous expanse and diversity of the world's forests do not lend themselves well to detailed description in a review like this. Moreover, while there is a long history of vegetation inventory and classification in temperate and some boreal forests, much less has been done in the tropical regions, outside of Australia. Fortunately, there are a number of excellent and comprehensive descriptions and summaries of regional forests; I have made extensive use of many of them here, and most are readily accessible. For readers interested in more detailed descriptions of regional floras, Frodin's (1984) guide to published floras is an excellent, though somewhat dated, reference.

The approach I have taken in classifying and describing the world's forests is primarily climatic-geographical-physiognomic in nature, with the forests separated into broad types based primarily on prevailing climate (e.g. tropical, temperate, Mediterranean, etc.). Within each type, I describe the regional formations, noting major species, cover types and physiognomic forms. The discussion of boreal and temperate forests contains the most specific references to individual species and forest types, as many of the species native to these regions are abundantly represented and widely distributed, and the regional floras are especially well known. I focus primarily on the naturally occurring vegetation in each region, especially mature forests.

I describe the forests of Australasia, eastern Asia, Europe, Fennoscandinavia and North America in the most detail, as these regions have been subject to the most research. They are also the regions for which the literature is most abundant and accessible, at least for an English-speaker like myself. I preface discussion of the regional formations by general information on the extent, distribution, prevailing climate and soils of each forest type. Throughout, I cite references to which interested readers can refer for more information about a region or forest type of particular interest to them.

## BOREAL FORESTS

Boreal forests are vast in size, widely distributed and remarkably variable despite a relatively low diversity of woody plants. Growing conditions range from extremely cold, low-moisture environments to perennially wet bogs and swamps of near-temperate climates. The trees are predominantly coniferous (especially *Abies, Picea* and *Larix* spp.), although a few deciduous genera *(Populus, Betula* and *Salix)* are nearly ubiquitous in their distribution and are often locally common. Vast areas of boreal forest remain relatively undisturbed by humans, although various disturbance agents, especially fire, have altered the composition and dynamics of these forests for millennia. Pressure from commercial harvesting is increasing, particularly in Siberian forests.

## Distribution and extent

Boreal forests, or taiga, are distributed globally across the high latitudes of the Northern Hemisphere (Fig. 2.1). Estimates of their extent range from 9 to 17 million km$^2$ (Lieth 1975; Perry 1994). The precise area is difficult to estimate, due to incomplete inventories and varying definitions of 'boreal forest', but the upper values (15-17 million km$^2$) are probably most accurate.

The northern limit of the North American boreal forest extends from about 68° N in Alaska to about 58° N at Hudson Bay and eastern Canada (Larsen 1980). It extends north to the Arctic Ocean in Scandinavia (above 70° N) and parts of western Russia and then follows a line ranging a few to several degrees above or below the Arctic Circle (66° 33' N). Boreal forests occur discontinuously southward to about 35° N at high elevations in the Appalachian Mountains of eastern North America and to similar latitudes in central Asia around the Indian subcontinent. Boreal forests grade into tundra at their northern edge, into montane coniferous, grasslands, parklands and mixed forests at their southern edges in North America and south-western Eurasia, and into woodland/steppe in the more central and eastern regions of Eurasia. The north-south width of the boreal forests ranges from about 500 to 2300 km in Eurasia and averages about 1000 km in North America (Larsen 1980; Archibold 1995).

## Climate

The climate of the boreal forest is characterized by low temperatures and relatively low precipitation. The growing season is short (90 or fewer frost-free days in northern areas), and some forests are underlain by permafrost. However, the long days during the growing season partially compensate for the comparatively unfavourable growing conditions.

The northern edge of the boreal forest corresponds roughly to a July isotherm of 13°C in both Canada and Eurasia (Larsen 1980). The southern border generally follows an 18°C July isotherm, except where precipitation is too low, as in the northern prairie region of western Canada. Winter temperatures can be exceedingly low, less than -70°C in parts of Alaska and Siberia. At the Bonanza Creek Long-term Ecological Research (LTER) site near Fairbanks,

Alaska, the mean daily temperature is -24.9°C, whereas the July mean is 16.4°C (Bonanza Creek LTER database). Mean annual temperature is 3.3°C, with an average of 233 days at below freezing temperatures annually. Summer temperatures in much of Canada and Eurasia are somewhat cooler (Larsen 1980).

Precipitation throughout the boreal region is typically less than in most temperate forests but is highly variable. In interior Alaska and Siberia, precipitation ranges from 100 to 270 mm (Rumney 1968; US National Oceanic and Atmospheric Administration records); in eastern Canada, however, it can be as great as 900 mm. Precipitation typically is greatest in summer, but where there is a maritime influence (e.g. eastern Norway and portions of Canada) it is distributed somewhat more evenly throughout the year and the autumnal drier periods characteristic of regions with a more continental climate are absent or less pronounced.

Common to all boreal forests are light regimes characterized by short winter and long summer days. Forests growing at or above the Arctic Circle experience at least 1 day each of both total light and total darkness. At the Bonanza Creek LTER site, average daily solar radiation ranges from a low of 750 $kJm^{-2}$ in January to 22 375$kJm^{-2}$ in June (Slaughter & Viereck 1986). Sun angles are low year round (49.5° on the summer solstice and 2.6° on the winter solstice in Fairbanks, Alaska) and accentuate the effects of topography. The low sun angle also dramatically increases the amount of twilight, which comprises an average of 10.4% of every 24h and about 17% of all light at 65° N compared with corresponding values of 3.1 and 6% at the equator.

## Soils

Spodosols, or podzols, are the dominant soils in boreal forests. Spodosols form as iron, aluminium and organic matter move from the A horizon into the B (spodic) horizon, a process known as podzolization. The word 'podzol' is of Russian origin and refers to the ashy-coloured, highly leached E horizon that lies between the A and B horizon(s). Spodosols are typically rather acidic, in the 3.5-4.5 pH range. Organic matter deposited around soil particles in the B horizon enhance cation exchange capacity, although

Spodosols are typically somewhat nutrient-poor. Because of the recalcitrant coniferous litter and harsh climate, decomposition is comparatively slow, and a thick litter horizon overlays the mineral soil in many upland environments. Moss is often abundant in the understorey, decomposing as much as 10 times more slowly than other plant litter.

Although dominant in area, Spodosols are by no means the only soils associated with boreal forests. Organic soils (Histosols) form in low areas where decomposition is extremely slow and the soil is wet all or most of the year. In the broad floodplains that occur along river corridors, Enti-sols and Inceptisols are often found. In montane environments, Alfisols are common (Archibold 1995), which have higher pH and base saturation than Spodosols.

Found in no other forest soils, permafrost is a permanently frozen layer of soil that underlies boreal forests growing on low, wet areas or on north-facing upland slopes. It forms when the soil does not warm sufficiently to fully melt during the summer, and may range from a few meters in thickness under forests to over 240 m under treeless tundra. Permafrost forms gradually, and is associated with an accumulating insulating cover of litter and especially moss on the forest floor during forest development. The active layer (the upper soil that melts during the summer) may be less than 1 m thick. *Picea mariana* and *Larix* spp. are among the few trees that grow over permafrost, and even they are of low stature (< 10-15 m). Where the active layer becomes too thin, the boreal forest or taiga gives way to tundra.

## Dominant forest types

Upland forests of various types dominate much of the boreal region. Common upland genera include the conifers *Picea, Abies* and *Pinus,* and the hardwoods *Populus, Betula, Sorbus* and *Alnus.* Deciduous boreal forests often have an open understorey, owing to their characteristically early place in succession and the fact that many of them establish after catastrophic fires. Coniferous forests are often associated with a denser under-storey and a well-developed moss layer.

Lowland communities include both forested and treeless communities. In addition to flood-plain terraces, swamps and muskeg may also be forested. Muskeg is a nutrient-poor environment, and wet-mesic to wet; it is an open environment and species-poor. Swamp is more nutrient-rich and supports a more diverse flora, including some deciduous species.

## Regional floras

### *Eurasian*

Ahti *et al.* (1968) divided the boreal region of Fennoscandinavia into four zones, arranged from south to north: the hemiboreal, southern boreal, middle boreal and northern boreal zones. The hemiboreal zone is dominated by broadleaf species of more temperate affinity, such as *Fraxinus excelsior, Quercus robur, Tilia cordata* and *Ulmus glabra.* The southern boreal zone is transitional between the hemiboreal and middle/ northern zones, and contains a mix of broadleaf and coniferous species. The middle and northern zones are largely coniferous, with *Betula* spp. the primary deciduous trees. The northern zone is characterized by a distinctly boreal flora (including abundant *Salix* spp.) and the absence of the low-herb spruce forests of the middle zone (Eseen *et al.* 1997). The conifers *Pinus sylvestris* and *Picea abies* are the dominant regional species. Both grow on a wide variety of sites, from xeric to wet-mesic. *Pinus sylvestris* tends to be more abundant on drier sites, which are more frequently disturbed by fire and have a more continental climate. Conversely, *Picea abies* grows on moister sites where fires are less frequent and a maritime climate often prevails.

Relative to parts of the North American and eastern Eurasian boreal forests, shrubs are infrequent in Fennoscandinavian forests. The woody understorey is more apt to comprise small trees and saplings. *Salix starkeana, Salix xerophila* and *Juniperus communis* occur, especially after disturbance, and *Sorbus aucuparia* and *Salix caprea* occupy mesic sites. The dwarf shrubs *Calluna vulgaris* and *Empetrum hermaphroditum* are common on dry sites, and *Vaccinium* spp. and *Ledum palustre* are widespread and may be locally common. Bryophytes and lichens are typically more abundant than vascular plants, and bryophytes often dominate the forest floor on mesic sites (Eseen *et al.* 1997). Two species of moss also common to the western North American boreal forests, *Holocomium splendens* and *Pleurozium schreberi,* are abundant on low-medium productivity sites. On dry sites and to the north, lichens become more important, especially *Cladonia* spp. and *Stereocaulon* spp. Various species and growth forms of *Cladonia* characterize early, mid and late successional communities (Ahti 1977; Oksanen & Ahti 1982; Oksanen 1983).

Deciduous species are more abundant in early and mid successional communities, and apparently occupied a much greater area of Fennoscandinavia prior to active fire suppression (Zachrisson 1985). Mature deciduous forests are important centres of biodiversity for a variety of life forms, including epiphytic non-vascular plants growing on *Populus tremula* and *Salix caprea* and numerous insects and several birds, including the white-backed woodpecker *(Picoides leucotos).*

The forests of western Russia are similar to those of Fennoscandinavia, with abundant *Picea abies* and *Pinus sylvestris,* along with *Picea obovata* and *Larix siberica.* While *Picea abies* does not grow eastward of the Kama River, *Pinus sylvestris* occurs throughout Eurasia (although its northern range becomes more restricted in the central and eastern regions). In Siberia, it forms nearly monospecific stands on low-nutrient sandy, rocky and boggy soils. In the mountains, it grows predominantly on eastern and western slopes, and will grow on permafrost where the active layer exceeds 2 m (Nikolov & Hemisaari 1992).

Numerous other species occur throughout the central and eastern portions of the Eurasian boreal forest. *Abies siberica* has a near-continental distribution, primarily below the Arctic Circle, and grows on productive upland sites (Nikolov & Hemisaari 1992). Common over-storey associates include *Picea obovata* and *Pinus siberica; Abies siberica* also occupies the under-storey of *Pinus sylvestris* and *Larix* spp. forests.

*Abies siberica, Betula pendula, Betula pubescens* and *Pinus siberica* grow on fertile upland sites in central Siberia, and *Betula pendula* forms early successional communities with *Populus tremula* after fire or other disturbance. *Picea obovata* also has a continental distribution, with a wider north-south range, and forms both pure (river floodplains) and mixed (upland) stands. It, *Abies siberica* and *Pinus sylvestris* are the principal species of the closed Siberian taiga forest (Nikolov & Hemisaari 1992).

In addition to the continentally distributed species described above, the far eastern boreal forest contains *Pinus pumila*, a shrubby species that grows in rocky mountain soils and in peat and alluvial soils in the lowlands. Common mountain associates include *Betula* spp., *Rhododendron* spp. and *Alnus fruticosa*, among others. *Pinus pumila* most often grows in the understorey of *Larix gmelinii (Larix duhurica)* forests, which typically form a monospecific canopy. *Larix gmelinii* is more widely distributed, growing in mountains, river valleys, coastal environments, uplands, at the forest-tundra border and at the southern extent of the boreal forests in forest-steppe communities (Nikolov & Hemisaari 1992). *Picea ajanensis* grows in both pure and mixed stands in the southeast boreal forest, occurring with *Larix gmelinii, Picea obovata* and *Betula* spp. in the overstorey on upland sites at high latitudes and elevations, and with a *Ledum* spp. and *Sphagnum* spp. understorey in low areas in the north-west of its range. In the more southerly portion of its range, it grows in mixed forests that variously include other *Abies* spp., *Acer* spp., *Betula* spp., *Fraxinus mandshurica, Juglans mandshurica, Populus* spp., *Tilia* spp. and genera of more temperate climates.

### *North American*

Larsen (1980) has extensively summarized the regional variation in the North American boreal forest, and I refer the reader to his text and the references contained therein for a detailed description. Viereck *et al.* (1992) have classified Alaska's forests in great detail, and their scheme is a useful reference for both Alaska and north-western Canada.

Although many species in the North American boreal forest are near continent-wide in their distribution, there is considerable regional variation in species associations and habitat preference (Table 2.1). In Alaska and Canada west of the Rocky Mountains, *Picea glauca* is the dominant late-successional species on upland sites devoid of permafrost and undisturbed by fire. *Populus tremuloides* and *Betula papyrifera* frequently colonize burned upland sites, and without further disturbance are succeeded by *Picea glauca* within about 100 years (Van Cleve *et al.* 1996). On flood-plain terraces, *Populus balsamifera* instead of *Populus tremuloides* is present, and *Betula papyrifera* is often less abundant (see Fig. 2.2).

Moreover, the formation of a dense, insulating moss layer (especially the feather mosses *Holo-comium splendens* and *Pleurozium schreberi)* leads to the formation of permafrost and the replacement of *Picea glauca* by *Picea mariana* on less frequently burned sites.

*Picea mariana* and occasionally *Larix laricina* dominate low wet areas, as well as north-facing slopes underlain by permafrost. These forests can degrade over time to open, wet communities dominated by low-stature ericaceous shrubs with few or no trees. However, fire can expose the mineral soil, resulting in permafrost melting and the re-establishment of forest vegetation. Further south, *Abies lasiocarpa, Picea engelmannii, Larix lyallii* and *Populus balsamifera* dominate the overstorey of high-elevation and subalpine forests.

In the boreal forests of the Mackenzie-Yukon region to the east, both *Picea glauca* and *Picea mariana* occupy the uplands, while the latter is more common along rivers and in lowlands. In west-central Canada, *Pinus banksiana* and *Pinus contorta* var. *latifolia* become common in the south, particularly on sandy soils of glacial origin and thin, coarse soils weathered from granitic outcrops. Both sandy and rocky soils share a number of species in common (Raup 1946; Thieret 1964). The ranges of *Pinus contorta* var. *latifolia* and *Pinus banksiana* overlap but the former is more abundant to the west.

In the Canadian Shield region to the east, *Picea mariana* dominates the thin soils common throughout this area. *Pinus banksiana* occupies drier sites, and *Larix laricina* is an associate on wetter sites, although open bogs and fens are common. Further east and south around the Great Lakes sub-boreal species at the northern edge of their range, including *Pinus resinosa, Pinus strobus, Thuja occidentalis, Acer rubrum, Acer saccharum* and *Betula alleghaniensis,* intermix with the prevailing boreal species. The Hudson Bay lowlands are largely wet, and are covered by open bogs and fens and open *Picea mariana* forests.

*Picea glauca* and *Abies balsamifera* are the most dominant upland species conifers in the central and southern regions of eastern Canada. Common hardwood associates include *Populus tremuloides, Populus balsamifera* and *Betula papyrifera.* Further north, *Picea mariana* covers most of the uplands, although *Pinus banksiana* occupies sandy sites and early successional *Picea mariana* sites. Forests near the forest-tundra transition are predominantly *Picea mariana-lichen* or *Picea mariana-Sphagnum* communities, while central, southern and coastal *Picea mariana* forests more commonly have a low shrub understorey. *Populus balasamifera* is common in Quebec, the Canadian Maritime Provinces and the north-eastern USA, particularly Maine (Eyre 1980); it is replaced by *Picea mariana* at higher elevations, and further west in Ontario comprises only a limited portion of the forest landscape.

## TROPICAL AND SUBTROPICAL FORESTS

The importance and fate of tropical and subtropical forests are topics of widespread and enduring interest. Long recognized as centres of high biotic diversity (perhaps containing more than half of the world's terrestrial plant and animal species), considerable effort has been expended on quantifying rates of deforestation, cataloguing species assemblages and identifying natural products of medicinal or other commercial importance. However, most of these efforts are of rather recent origin and tropical forests remain perhaps the least understood of all terrestrial ecosystems.

### Distribution and extent

The tropics proper lie between the Tropic of Cancer (23.5°N) and the Tropic of Capricorn (23.5°S). About 23-35° north and south of these latitudes are regions similar in climate and vegetation type and form, commonly referred to as subtropical. There are perhaps 28-32 million km$^2$ of (sub)tropical forest, about 60% of which is dry or seasonal forest, the remainder being humid or wet.

Tropical forests are further defined by their location within one of three geographical formations: American (or Neotropical), African and Indo-Malesian. Within the American formation, forests are most abundant in the Amazon drainage basin. Brazil itself contains 352 million ha of intact moist tropical forest, 16 million ha of other closed canopy forests, 155 million ha of intact woodlands and about 35 million ha of degraded secondary forests of various types (Schroeder & Winjum 1995). Other areas with extensive forests are north-west South America, Central America and Mexico. Numerous islands in the Caribbean Sea have, or did have, (subtropical forests, and forests along the coast of the Gulf of Mexico in the USA are climatically and floristi-cally subtropical, particularly in Florida and south Texas. In total, the American formation comprises about 50% of the world's tropical forests.

## Climate

The tropics generally correspond to frost-free areas, especially those where the mean annual temperature is 18°C or above (Walter 1985) or where the mean monthly temperature difference between the three warmest and three coldest months is 5°C or less. Together the tropical and subtropical regions comprise about 58% of the earth's surface area (Kellman & Tackaberry 1997).

Precipitation throughout the tropics is highly variable. Most rainfall is convective in origin, of short duration and high intensity, and patchily distributed (Kellman & Tackaberry 1997). At least 2000 mm of annual precipitation is generally regarded as necessary for the occurrence of rain forests (Lauer 1989). Of even greater importance, however, is that rainfall is distributed rather evenly throughout the year. Thus, rain forests occur in portions of Africa where annual precipitation is only 1600 mm but falls throughout the year. In areas where there are dry (<100mm) periods of a few to several months in duration, and as precipitation declines to 1000 mm or less annually, rain forests are replaced by dry forest or savannah.

## Soils

Most soils in the tropics are old and highly weathered. Dominant lowland soils are Oxisols and Ultisols, which together comprise about 23% and 18% of all non-desert tropical soils respectively (Richter & Babbar 1991). Other common soils include Alfisols (10.5%), Vertisols (3.4%) and young soils of volcanic or alluvial origin (22.4%). Mountain soils of various orders comprise about 18% of tropical soils, while organic soils occur on less than 1% of the landscape.

Oxisols form in warm, wet environments and are commonly associated with tropical rain forests. There is typically little horizon differentiation. Iron and aluminium oxides are abundant, while potassium, magnesium and available phosphorus are low. Both pH (4.5-5.5) and cation exchange capacity (<16mEq per 100g of clay) are low. Large concentrations of Oxisols are found in the Amazon and Congo basins. Ultisols are typically associated with dry tropical forests and develop under seasonal precipitation regimes. Ultisols have a distinctive argillic horizon with a high concentration of iron sesquioxides. Organic matter content and cation exchange capacity are typically low, with a base saturation of less than 35% in the lower horizons. Ultisols are common in south-eastern North America and South-East Asia, although both Oxisols and Ultisols frequently occur together, along with one or more of the other dominant soil orders.

## Dominant forest types

Although the term 'tropical forest' is often used as a surrogate for tropical rain forest, there are in fact a variety of regional forest types. Longman and Jenik (1987) identified five primary forest types: tropical evergreen, (sub)tropical seasonal, (sub)tropical semideciduous, sub(tropical) evergreen and mangrove.

Tropical evergreen forests (i.e. rain forests) include:

1 lowland subtypes, characterized by tall emergent trees (>30m) projecting above a multilayered canopy, with sparse undergrowth and few epiphytes;

2 montane forests with lower stature trees and a rich woody understorey and herbaceous ground cover;

3 cloud forests with a closed canopy with gaps, low stature (<20m) canopy trees, abundant lianas and climbers, and a rich herb, moss and fern cover;

4 alluvial multilayered forests (sometimes referred to as gallery forests) growing in flood-prone environments, with abundant herbaceous vegetation and epiphytes;

5 swamp forests growing in frequently or perennially flooded areas with abundant ferns and herbs;

6 peat forests with low-stature trees, poorly developed ground cover and occasionally abundant palms.

After disturbance, tropical evergreen forests frequently develop a thick undergrowth. The (sub)tropical seasonal forests are predominantly evergreen, with some trees of deciduous habit.

They range from lowland to montane environments, with climbers and evergreen shrubs present and tree ferns present at lower elevations. The (sub)tropical semideciduous forests are dominated by overstorey trees that lose their foliage during the dry season, although the understorey is often evergreen. Ground cover is abundant but epiphytes are not. The (sub)tropical evergreen forests are structurally similar to tropical evergreen forests, but less productive due to periods of colder temperatures during the winter. Finally, mangrove forests are found in tidal areas, are evergreen in habit, are adapted to the salt-water environment, and have an understorey nearly devoid of vegetation but often containing non-vascular plants.

## Regional formations

### *American*

The forests of the Neotropics are exceptionally diverse, particularly the rain forests, and their phytogeography is complex. Archibold (1995) summarized Longman and Jenik's (1987) and Mabberly's (1992) data on the most significant woody plant families in the region, which include the Euphorbiaceae, Lecythidaceae, Leguminosae, Meliaceae, Moraceae, Myristicaceae and Sapotaceae. Prance (1989) identified 18 regions or flor-istic provinces within South American tropical forests, based upon species distribution and areas of high endemism; he defined four broad categories, into which the 18 provinces could be classified. Of perhaps more general usefulness is a classification scheme based on climate, topography and vegetation type: (sub)tropical moist or broadleaf evergreen (i.e. rain forests), montane broadleaf, wetland, and sea-sonal/(semi)deciduous.

Moist (sub)topical forests are diverse, multi-storeyed communities dominated by large-stature trees. Vines and other climbers are often abundant. These are low-elevation forests, growing in areas of high year-round rainfall. The largest concentration of these forests is in the Amazon basin (including Guyana, Surinam and French Guiana), although they occur in five other principal areas (Prance 1989): (i) Mexico and Central America (predominantly along the Gulf of Mexico and along the Caribbean), (ii) several Caribbean islands (Cuba, Puerto Rico, Hispaniola and

Jamaica and some of the Lesser Antilles), (iii) the Pacific coast of South America (northern Colombia to northern Ecuador), (iv) northern Venezuela and (v) the Atlantic coast of Brazil. Northern Venezuela, Amazonia and associated areas and both coastal regions are characterized by high levels of endemism, while the other two regions have mixtures of both endemic and widely distributed species.

Montane forests occur in both humid and arid areas, and the associated forest types are consequently quite variable. Subtypes include lower montane and upper montane and subalpine forests (Prance 1989). Andean lower montane forests begin at 700-1200 m and range up to 1800-2400 m in altitude. Species in the genera *Licania* (Chrysobalanaceae) and *Eschweilera* (Lecythidaceae) occur here as well in the lowland forest. In northern Peru, podocarp (especially *Podocarpus oleifolius)* forests occur. Situated between the lower montane and subalpine forests (beginning at 3400 m or less), the Andean montane forests comprise more distinctive high-elevation species, including *Brinellia occiden-talis, Symplocos pichindensis* and *Weinmannia balbisiana*. Subalpine forests may grow at elevations up to 3800m and are dominated by low-stature species in families such as Ericaceae and Rhamnaceae. Cloud forests are characterized by particularly dense canopies and dark under-stories, abundant epiphytes (orchids, bromeliads, ferns, mosses and lycopods), and abundant shrubs but sparse vines. These forests are common along the eastern slope of the Andes, and are found in south-eastern Brazil, high elevations in the Greater and Lesser Antilles, and parts of Central America (chiefly Guatemala, Panama, Costa Rica and Nicaragua). In Central America, the montane flora is more similar to that of southern North America, including species in the genera *Quercus* (Fagaceae) and *Liquidambar* (Hamamel-idaceae) (see p. 49).

American tropical wetland forests include permanently flooded forests, periodically flooded forests and gallery forests. Permanently flooded forests are often dominated by palms *(Mauritia* and *Euterpe)*. Periodically flooded forests include mangrove forests, which grow in marine tidal areas, and various freshwater and tidewater swamp forests. Mangrove forests occur from southern Florida southward along both the Atlantic and

Pacific coasts of Central and South America. Eight species of mangrove occur in the American formation, among which *Rhizophora mangle, Avicennia germinans, Laguncularia racemose* and *Conocarpus erecta* are common (Kricher 1997). Mangrove forests are typically low in stature, although *Avicennia germinans* may reach heights of 20 m. Pacific coast mangrove forests are typically more diverse than their Atlantic counterparts. Detailed descriptions and comparisons of mangroves and mangrove forests can be found in Pool *et al.* (1977), Tomlinson (1986) and West (1977). Tropical tidal swamp forests are most abundant on the Amazon river delta, and are rather diverse. Palms (e.g. *Astro-caryum murumuru, Euterpe oleracae, Jessenia bataua, Manicaria saccifera, Mauritia flexuosa, Maximiliana regia, Oenocarpus* spp. and *Raphia taedigera)* are common (Prance 1989). Further descriptions of these forests can be found in Anderson and Mori (1967).

Seasonal and semideciduous forests occur where there is a significant dry season and are found throughout the Neotropics. Significant areas occur from north-western Mexico to north-western Costa Rica (Kricher 1997). They also grow in Amazonia, southern Brazil, Venezuela, Paraguay and Bolivia. Rates of endemism are typically low, although there are significant geographical differences in species composition. Vast areas that formerly supported seasonal forests have been converted to the cultivation of coffee *(Coffea arabica)*. The most abundantly woody plant families in Neotropical dry forests are generally Leguminosae and Bignoniaceae. Mytaceae is dominant in some forests of the Caribbean West Indies and often abundant elsewhere (Gentry 1995). Also common are representatives of Rubiaceae, Sapindaceae and Euphorbiaceae.

### *African*

Aubrévillé (1956) recognized three primary regions of African forest: the Guinea Forest, the Nigerian Forest and the Equatorial Forest. The Guinea Forest extends about 1280 km from the Guinea coast and Sierra Leone eastward across Liberia, Ivory Coast and Ghana. Although the forests of Nigeria become more species-poor from east to west, the floristic distinction between the Nigerian and Equatorial forests is no longer recognized as significant enough to warrant separation between the two (White 1983). However, the savannahs (now largely agricultural) in Togo and Benin form a natural floristic boundary between Aubréville's Guinea Forest and the forests to the east. This area extends over the Nigerian lowland to Cameroon (Aubréville's Nigerian), and from Cameroon eastward over 2400 km into the African interior and the Ugandan plateau (Equatorial). Most of the forested area lies in the Congo Basin of northern Zaire (Democratic Republic of Congo) (Archibold 1995). Monod (1957, in Hamilton 1989) suggested that the rain forests of Africa comprise one large Guineo-Congolian Region. He further proposed that this region can be divided into two groups (the west-central Atlantic-Congo Domain and the eastern Oriental Domain), because the rain forests of western Africa are floristically distinct from the central and eastern forests.

African forests bear some taxonomic similarity to the American formation. For example, legumes (Leguminosae) are abundant, particularly in dry forests. Other families found in both formations include Euphorbiaceae, Meliaceae, Moraceae and Sapotaceae, all of which are believed to be old families that occurred on both continents prior to the break-up of Pangaea (Hamilton 1989). Generally, lowland tree species richness is much less in African forests than in either the American or Indo-Malesian formations, and single species often dominate the canopy. These include *Brachystegia laurentii, Gilbertiodendron dew-evrei* and *Cynometra alexandri* (Hamilton 1989). However, there are generally recognized centres of floristic diversity (and, typically, endemism) within Africa proper. These occur in Cameroon and Gabon; eastern Zaire (Democratic Republic of Congo); eastern Ivory Coast and western Ghana; Sierra Leone and Liberia; and the vicinity of the East African coast (Hamilton 1989). The areas of high floristic diversity are associated with high rainfall, but even more importantly these regions probably had mesic climates during severe Quaternary climatic fluctuations that caused widespread extinctions in other areas during very arid periods. Patterns of species richness in montane forests are somewhat different, most notably in that montane forests in and around Cameroon lack much endemism, whereas there is a high degree of endemism in lowland forests in the area. Forests are floristically diverse in the Eastern Arc Mountains of Tanzania, and species change progressively with elevation (Lovett 1996). However, community diversity is not strongly correlated to elevation or latitude; instead, disturbance, low precipitation and a pronounced seasonal climate all reduce species richness.

The forests of southern Africa are variously referred to as either subtropical or temperate, but here I consider them subtropical due to their taxonomic similarity with Malesian forests, largely evergreen habit and low-latitude occurrence. The species assemblages at higher elevations and along the south coast resemble those of the montane forests of East Africa. *Podocarpus* spp. and *Olea laurifolia* are common, as is *Apodytes dimidiata, Curtesia dentata* and *Ocotea bullata* (Adamson 1956). Shade-tolerant shrubs and ferns occupy the understorey. Wetter montane forests contain *Cunonia capensis, Ilex mitis* and *Platylo-phus trifoliatus.* The endemic *Widdringtonia* (including *W. juniperoides* and *W. scrrwarzii)* forests grow in small areas north of Cape Town.

On the south-west coast are forests of *Podocar-pus henkelii, Peroxylon utile* and *Xymalos monosperma* and the shrubby *Buxus maccrwani* (Adamson 1956). The east coast forests of Natal further to the north are more species-rich than those of the south coast. Podocarps are again common, and among the associated species are *Calodensrum capense, Celtis africana, Combre-tum krausii, Olea capensis, Ptaeroxylon obli-quum* and *Zanthoxylu dayvi.*

### *Indo-Malesian*

The Indo-Malesian tropical forest formation is the second largest of the three and has been extensively studied. Geographically, the formation ranges from India through Polynesia, and includes the east coast of Australia and the South-East Asian mainland. The regional forests are diverse in both type and composition, and include rain forests, monsoon forests and savannahs among others. The wide array of landforms, climates, soils and the degree and types of human activity within the region interact to add further layers of floristic and structural complexity. Regional floristic diversity is extremely high. About half of all plant families (220) and one-quarter of all genera (2400) are represented just within Malesia (the region from Sumatra and Malaya eastward to the Bismarck archipelago, north to the

Malaysia-Thailand border and south to between New Guinea and Australia) and there are 8000-10000 species of small to large trees (Whitmore 1989). In Malaysia and Singapore, there are 82 families and 388 genera of trees; 27% of the tree species are endemic (Ng & Low 1982). Indo-Malesian tropical forests differ taxonomically from the American and African formations in that a number of conifers are well represented. Two families comprise most coniferous species and genera, the Araucariaceae and Podocarpaceae, although *Pinus* extends into western Malesian monsoon forests (Whitmore 1989).

The monsoons are the dominant climatic factor governing precipitation in the region, and hence determine the occurrence and distribution of evergreen and seasonal forests. Moist evergreen forests are centred in the Malay archipelago (Whitmore 1989). Whitmore (1975, 1984) identified 13 types of rain forest (both evergreen and semi-evergreen) in the region, differing primarily in the type of sites they occupy (lowland, swamp, montane, etc.). Moist evergreen and semi-evergreen forests occur throughout Malaysia and Indonesia, although extensive logging and land-use changes have significantly reduced their extent. Dominant families include Anacardi-aceae, Apocynaceae, Burseraceae, Bignoniaceae, Lecythidaceae, Leguminosae, Meliaceae and Rhamnaceae, among others (Sewandono 1956). Dipterocarpaceae are common in the western portion of Malesia on the landmasses of the Sunda Shelf (especially Sumatra, Malaya, Borneo, Palawan and Java), which projects out from the Asian mainland. Groups of dipterocarps may reach canopy heights of 60 m. Representatives of Fagaceae are common in montane forests.

Rain forests extend northward through the mainland and into China and a more subtropical climate, where common tree genera include *Quercus, Diospyros, Machilus, Castanopsis* and *Cinnamomum camphora* (see p. 40). The inland plains of Burma (Myanmar), Thailand, Cambodia and Laos lie in the rain shadow of the mountains of western Burma, and drier forests are more common. In the south, dipterocarps and legumes predominate at lower elevations, with dipterocarps and *Quercus* spp. at high elevations. Monsoon forests are scattered throughout, especially in south-eastern Indonesia and portions of New Guinea, but are greatly reduced in extent owing to human disturbance. In western India, moist evergreen forests occupy the coastal region from about 21° S to near the southern tip of the peninsula. The Myrtaceae and Lauraceae are well represented, as are dipterocarps and species in the genus *Hopea*. Rain forests also occur in western Pakistan and eastern Burma. Inland of the west Indian coastal rain forest lie the monsoon forests, which are dominated by a deciduous overstorey and evergreen understorey. Important timber species in the region include teak *(Tectona grandis)* and rosewood *(Dalbergia latifolia)* (Chaturvedi 1956). Larger areas of Indian monsoon forest occur in the eastern central region and in the foothills of the Himalayas. Xerophytic forests dominate northwestern India and the western peninsula (inland of the rain forests and monsoon forests). Various species of *Acacia* are common throughout. The remainder of the forests in and around India are predominantly dry deciduous forests. Teak is present in places, and in southern India sandal-wood *(Santalum album)* is one of the most important timber species.

Australian tropical rain forests occur predominantly along the north-eastern (Queensland) coast and occupy less than 1% of the continent (Webb & Tracey 1994). Canopy trees include conifers in the Araucariaceae and Podocarpaceae and at least 33 dicotyledonous families. Among the better represented are Elaeocarpaceae, Lauraceae, Moraceae, Myrtaceae, Proteaceae, Sapotaceae and Sterculiaceae (Beadle 1981). Vines are characteristically abundant, while tree ferns and palm are occasionally present lower in the canopy. Monsoon, or semi-evergreen, forests are typically small in size (1-25 ha) and occur in the Arnhem peninsula, Melville Island and portions of Western Australia (Beadle 1981). Canopy species found in these forests vary geographically and by moisture regime, but include *Bombax ceiba, Buchaniana arborescens, Canarium aus-tralianum, Gmelia dakympleana* and various species of *Ficus*. Common shrub and small tree genera include *Canthium, Croton, Diospyros, Exocarpus, Litsea, Livistona, Myristica, Pongamia* and *Trema*. Many of the northern Australian dry forests contain genera characteristic of moist tropical forests and are believed to be derivatives of former rain forests that occupied the area during a period of more favourable climate.

Most of tropical Australia receives insufficient rain to support rain forest or monsoon forest and is instead dominated by *Eucalyptus* forest and woodland. Along the eastern coast, intermixed with the moist tropical forests previously described, are tall *Eucalyptus* forests that grow on fertile soils. Canopy species are occasionally found in the moist tropical forests, and rain forest species are commonly found in the understorey of the tall *Eucalyptus* forests (Beadle 1981). Low-stature (sub)tropical *Eucalyptus* forests and woodlands occur on nutrient-poor soils of the north-eastern coast, the northern portion of the eastern inland lowlands, and the north central region. *Eucalyptus* forests of varying stature also occupy the subtropical south-western corner of Australia. Mallee, woodlands dominated by multistemmed shrubby *Eucalyptus*, occur primarily in the south-west and along the south-western coast. In drier interior regions, *Eucalyptus* is replaced by woodlands typically dominated by various species of *Acacia* or *Casuarina*.

## TEMPERATE BROADLEAF AND CONIFEROUS FORESTS

Of closed canopy forests, perhaps none has been as extensively utilized and altered by humans as the temperate forests. Significant portions of Europe and eastern Asia that once supported forests have long since been converted to pasture and agriculture, and little of the original vegetation remains. Similarly, much of the temperate forest in the USA has been cut, although a large percentage returned to forest after cutting or agricultural abandonment. None the less, forests still occupy moderate to significant portions of the temperate regions of North and South America, Europe, Asia and Australia, and are of significant economic and ecological importance.

### Distribution and extent

There are about 1.1 million km$^2$ of temperate broadleaf deciduous forests in the world, with perhaps another 3-8 million km$^2$ of temperate coniferous and broadleaf evergreen forests and woodlands. In North America, temperate broadleaf forests are confined primarily to the eastern USA and south-east Canada. West of the central prairie region are the interior and mountainous coniferous forests, and beyond them the coastal temperate coniferous rain forests of the Pacific North-west. Latitudinally, they occur between the subtropical zone (about 28-30°N) and the boreal zone (46-47° N), although the temperate coniferous rain forests of the Pacific coast of Canada and Alaska grow a little north of 56° N. Smaller areas of temperate forest occur in southern Mexico and into Belize and Guatemala, principally at high elevations and along river courses at 0-2500 m (Röhrig 1991a). The South American temperate forests grow primarily along the far south-western coast and the southernmost tip of the continent. The deciduous forests east of the Andes grow south of 45° S, while the coastal rain forests grow between 41 and 56° S. Temperate broadleaf evergreen forests grow principally in Chile, with a minor extension into Argentina.

European temperate forests occur predominantly in western and central Europe, continuing in a narrow band through eastern Europe to the Ural Mountains in Russia. Temperate deciduous forests also occur in the Near East in a band around the southern Caspian Sea, broadening to the east into the Iranian highlands, Turkmenistan, Uzbekistan and Kazakhstan (Röhrig 1991b). Mountainous areas are forested primarily by conifers.

Asian temperate deciduous forests are found primarily in the central and south-eastern region, including mainland China, Taiwan, Korea and Japan. The region lies between about 26° and 50° N, with conifers increasing in importance with latitude and elevation. Temperate forests, primarily coniferous but often with abundant deciduous canopy species, also occur in central Asia, principally at higher elevations in the Himalayan region. In Australia, moist temperate deciduous forests are found predominantly in northern and western Tasmania, while the mainland is occupied principally by the evergreen *Eucalyptus* forests of the south-east and southwest. Much of New Zealand is temperate broadleaf evergreen forest.

## Climate

As expected for a forest type distributed so widely, prevailing climatic conditions vary substantially. Generally, though, temperate environments are characterized by warm, relatively humid summers and cool-cold winters. Where precipitation is insufficient to support closed canopy forests, temperate forests variously grade into grassland (steppe or prairie), savannah or open woodlands. To the north, where both temperature and precipitation decline, they merge with the boreal forests, and to the south with (sub)tropical vegetation.

At lower elevations, mean temperature during the coldest month typically ranges between -5 and 10°C, with mean temperatures between 10 and 18°C. In mountainous regions, the temperatures can be significantly lower, approaching those of the more northerly boreal zone. Mean annual temperatures typically range between 8 and 13°C but can be lower or higher, depending on geographical location and local elevation (Röhrig 1991c).

Precipitation is year-round, although there are significant geographical differences in seasonal distribution. Winters are wetter than summers in much of the temperate forest region, although precipitation is more evenly distributed in portions of eastern North America, western Europe, far southern South America and New Zealand. In the temperate coniferous rain forests of the north-western USA and southeastern Canada, summers are dry and most precipitation falls during the late autumn and winter. Mean annual precipitation is less than 500 mm in some portions of South America and Russia. At the other extreme are the temperate coniferous rain forests of the north-west North American Pacific coast (cover photo), where annual precipitation approaches 2500 mm. More typical values for the temperate forests of eastern North America, western Europe and South-East Asia are 550-1300 mm (Röhrig 1991c; Archibold 1995).

## Soils

Soils are naturally variable, although the dominant soil orders found under broadleaf temperate forests are Alfisols, Inceptisols and Ultisols, with substantial areas of Entisols in floodplain environments (Archibold 1995). Spodosols and Ultisols are most common under temperate coniferous forest, but ash-derived Andosols are common in volcanic regions. Alfisols, Inceptisols and some Ultisols are relatively fertile, with high base saturation. Soil pH typically ranges from slightly acidic to slightly basic, although under conifers and some hardwoods, or in low-lying areas with impeded drainage and decomposition, soil pH may be 4 or less.

## Regional formations

Although there are distinct regional compositional differences, there is remarkable taxonomic similarity among the Asian, European and North American formations. Numerous genera are common to all three, including *Acer, Betula, Fagus, Fraxinus, Populus, Quercus* and *Ulmus*. The flora of North America and eastern Asia are most similar to one another, and share several genera of forest trees not found in Europe, including *Catalpa, Diospyros, Liriodendron, Nyssa* and *Sassafras* (Röhrig 1991c). There are further such affinities in the herbaceous flora of the forest floor (Li 1972).

Although the modern floras evolved under similar climatic regimes, the taxonomic similarity is due primarily to their common geological past. Both the North American and Eurasian continents were connected until the mid-Tertiary, after the period (Cretaceous) during which most of the genera common to the temperate forests of the Northern Hemisphere differentiated. The South American and Australasian formations do not share a common geological past and differ tax-onomically from those of the Northern Hemisphere. However, they do share some families (e.g. Fagaceae). Of the three Northern Hemisphere formations, the Asian is the richest, followed by the North American and the European distantly third. The comparatively depauperate European temperate flora is believed to have resulted largely from the east-west orientation of the region's mountain ranges, which inhibited migration and reduced available refugia during past glaciations.

### *Asian*

On the Asian mainland, temperate deciduous forests occupy a coastal band along the China Sea from the Korean Peninsula and into China in the vicinity of Tianjin. From here, deciduous forests broaden throughout the Chinese lowlands to the west (to the Mongshan Mountains) and south to the Yangtze River. Intermixed along the Chinese coast and in the interior highlands are mixed broadleaf-coniferous forests. Further to the south, evergreen broadleaf trees become more abundant. Coniferous forests become more prominent in the northern parts of both China and Korea.

Japan's forests consist primarily of temperate broadleaf deciduous communities in the central-north region, with conifers more abundant at higher elevations and latitudes. To the south, between about 26° 30' and 35° 30'N, lie the temperate broadleaf evergreen forests (Satoo 1983). Ching (1991) has described the composition of the region's forest in some detail and I only briefly summarize them here.

*China.* In China's north-eastern temperate forest region (eastern Heilongjiang and Liaoning provinces), conifers occur at the highest elevations and the northern latitudes, lending the forests a (sub)boreal character. Below these forests (up to 1100m elevation) are the *Pinus koraiensis-mixed* hardwood forests, some of which are rather species-rich. Subtypes include the mixed hardwood-*Pinus koraiensis* forest, oak forests *(Quercus mongolica)* and *Betula* forests, the latter of which occur at higher latitudes and elevations (Ching 1991).

China's northern temperate forest is roughly triangular in shape, extending from 32° 30' to 42° 30'N and from 103° 30' to 124° 10'E (Ching 1991). The major types within the region include the northern oak forest and the southern oak forest. Common throughout the northern oak forest is *Quercus liaotungensis,* with increasing *Quercus mongolica* to the east. The Liaotung hills and plains comprise the central portion of the northern oak forest area, with forest composition varying by landform and geography. In the mountainous regions of the Hebei and Laioning hills, broadleaf trees grow to elevations of about 1600 m, the forests being dominated by *Quercus* spp. along with *Betula* and *Populus.* The forest of the high loess plateaus of Shaanxi and Shanxi consist of *Pinus* and various hardwoods. The *Picea meyeri* and *Picea wilsonii* characteristic of the original primary forest have been widely replaced on the plateau by *Betula* in disturbed areas (Ching 1991).

The southern oak forest of the northern region includes not only deciduous species but also broadleaf evergreens. It is distinguished from the northern oak forest by the scarcity of *Quercus liaotungensis* and the presence of the evergreen *Quercus variabilis.* The vegetation of eastern Shandong bears close similarity to many of Japan's forests and is floristically diverse. Frost-tolerant *Tilia* spp. and *Corylus* spp. are found to the north, and *Machilus thunbergii* is present further south. The central and southern Shandong hills are forested with *Quercus*-dominated communities, as are the plains and hills of southern Shanxi and central Shaanxi. *Betula* and *Salix* become progressively more important with elevation, as do the conifers *Pinus armandii, Pinus tabulaeformis* and *Abies fargesii* (Ching 1991).

The Yangtze River drainage in east central China is a transitional zone between the temperate forests to the north and the subtropical forests to the south. There is a high degree of endemism. The genus *Quercus* is well represented, and the tree flora is a mix of more northerly and southern species. Deciduous species and conifers are intermixed with the broadleaf evergreens, which include the genera *Actinodaphne, Castanopsis* and *Lindera.* Bamboo 'forests' grow here, including *Aruninaria* spp. and *Phyllostachus* spp. The upper river valley is particularly diverse, and includes 50 broadleaf and 12 coniferous genera.

Forest of *Abies, Picea* and *Tsuga* occupy the high elevations.

*Korean Peninsula.* There is disagreement over the classification and latitudinal zonation of Korea's forests (Hyun 1956; Yim 1977a,b; Ching 1991) but the following types are relatively distinctive: warm temperate broadleaf evergreen, southern deciduous, northern deciduous and cool temperate/boreal coniferous.

The warm temperate broadleaf evergreen forests occupy the southernmost coast of the Korean Peninsula, and include *Quercus* spp., *Castanopsis* spp., *Cinnamomum camphora* and *Machilus thunbergii* (Hyun 1956). Temperate deciduous forests lie north of approximately 35° N and account for about 85% of Korea's forested land base (Hyun 1956). Species common to the southern deciduous forests are *Acer formo-sum, Carpinus laxiflora, Carpinus tschonoskii, Quercus mongolica* and *Quercus serrata.* The northern deciduous species include *Celtis koraiensis, Choseia bractesa, Hemiptera dividii, Larix olgensis, Magnolia parviflora* and *Populus maximowiczi.* Other species found variously in the temperate region include *Acer mono, Betula* spp., *Carpinus* spp., *Fraxinus* spp. and numerous *Quercus* spp. In the plateaus and mountains of northern Korea, coniferous species dominate the overstorey, including *Abies nephrolepis, Abies holophylla, Larix dahurica* and *Picea jezoensis* (Hyun 1956; Archibold 1995). *Pinus pumila,* a dwarf species, grows at the highest elevations (above 1200-1600 m). Broadleaf understorey species include *Acer* spp., *Betula* spp., *Tilia* spp. and *Ulmus* spp.

*Japan.* The forests of Japan include warm temperate broadleaf evergreen, temperate deciduous and cool/cold coniferous. The warm temperate broadleaf evergreen forests occur on the islands of Honshu, Shikoku and Kyushu, and extend from the lowlands to 600-850m (Satoo 1983). Species characteristic of the region include *Castanopsis cuspidata, Cycobalanus* spp., *Cinnamomum* spp., *Machilus* spp. and *Quercus* spp. In more mature forests, *Quercus gilva* grows on deeper soils, *Quercus slicina* on steep slopes and *Quercus glauca* over limestone (Archibold 1995). Following disturbance, *Quercus acutissima, Quercus serrata* and *Quercus variabilis* establish in mid-succession. The conifers *Abies firma, Podocarpus macrophyllus, Podocarpus nagi, Torreya nucifera, Pinus thunbergii* and *Tsuga sieboldii* variously occur, depending on latitude and proximity to the coast.

Temperate deciduous forests on the four main Japanese islands (Hokkaido, Honshu, Shikoku and Kyushu) are diverse in species and habitat, too much so to adequately describe here. In general, though, *Fagus crenata* dominates many forests. Its associates vary, depending on topography, disturbance history, soils, etc. Sites facing the Sea of Japan are commonly cold, snowy and windy, and species like *Acer tschonoskii, Acer matsumarae, Alnus maximowiczii, Quercus mongolica* and *Sorbus commixta* are common (Ching 1991).

Various species of *Alnus, Populus* and *Salix* grow in floodplain environments, and the conifers *Abies sachalinensis, Abies firma, Abies mariesii, Picea jezoensis, Pinus thunbergii* and *Tsuga sieboldii* variously grow with species of *Betula, Carpinus, Fraxinus, Magnolia, Quercus, Tilia* and *Ulmus* in montane regions. *Quercus cripsula* is also a common dominant, particularly on Hokkaido, and grows well on somewhat drier sites than *Fagus crenata.* Its associates include *Abies sachalinensis, Acer mono, Betula ermanii, Kalopanax pictus, Ostrya japonica, Picea jezoensis, Tilia japonica* and *Ulmus davidiana* var. *japonica* (Ching 1991). *Pinus densiflora* sometimes forms pure stands, with a broadleaf understorey. At higher latitudes and altitudes, the cool/cold coniferous forests are dominated by *Picea* spp. and *Abies* spp. Broadleaf associates include

*Betula* spp., *Fraxinus mandshurica, Kalopanax septemlobus* and *Fagus crenata.*

### *Australasian*

*Australia and Tasmania.* The majority of Australia's forests are considered (sub)tropical, some of which (the south-western and eastern *Eucalyptus* forests) have been described above (see p. 36). Webb and Tracey (1994) consider Australian forests south of about 38° S to be temperate, and I adopt that latitudinal boundary here. However, others consider the temperate forest zone to extend as far north as about 25° S (Bailey 1989), and in reality the transition from (sub)tropical to temperate is characteristically indistinct.

Regardless of the precise line of demarcation, the temperate forests of the Australian mainland are dominated primarily by *Eucalyptus,* and account for approximately 29 million ha of a total 42 million ha of closed forest, both temperate and tropical (Ovington & Pryor 1983). Groves (1994) has described the dominant vegetation in some wet sclerophyll Australian temperate *Eucalyptus* forests distributed among four main geographical regions (Table 2.4). Similar tabulations, including both wet and dry sclerophyllous *Eucalyptus* forests divided into major alliances, have been made by Ovington and Pryor (1983). Despite the diversity of *Eucalyptus* (>450 species), temperate forests are often dominated by only a few species, for example *E. obliqua-E. viminalis* forests in parts of Victoria and *E. marginata-E. calophylla* forests in Western Australia (Ovington & Pryor 1983).

Temperate rain forest occupies small areas of south-eastern Australia, principally in the territories of Victoria, the north-eastern corner of New South Wales and far south-eastern Queensland. The canopies of Victorian forests are dominated by *Nothofagus cunninghamii,* while *Athersperma moschatum, Acacia melanoxylon, Eucriphia lucida* and *Phyllocladus asplenfolius* are occasional or common associates (Ovington & Pryor 1983). *Nothofagus cunninghamii* does not grow in easternmost Victoria, and *Athersperma moschatum, Eleaocarpus holopetalus* and *Telopea oreades* are most common (Busby & Brown 1994). Eucalypts dominate the drier uplands in the region (Howard & Ashton 1973; Ashton & Attiwill 1994). In the forests of New South Wales, *Eucryphia moorei,*

*Nothofagus moorei* and *Doryphora sassafras* are common, with the former two species occurring at somewhat higher elevations.

Western Tasmanian temperate rain forests are dominated by *Nothofagus cunninghamii*, which may contain any of about 70 associates, only a few of which typically occur together in a stand (Ovington & Pryor 1983). *Athrotaxis* spp. (especially *A. cupressoides)* occur at altitudes greater than 900 m. In drier environments, like eastern Tasmania, *Eucalyptus* spp. are dominant; common ones include *E. delegatensis, E. dalrym-pleana, E. gunnii, E. archeri, E. coccifera, E. urnig-era* and *E. pauciflora.*

*New Zealand.* The climate of New Zealand ranges from subtropical to cool/cold temperate, but here I consider it mainly temperate. Generally, the native taxa are similar to those of the South American and Tasmanian forests. Most are evergreen in habit, including both conifers and broadleaf species. There are two broad categories of New Zealand forest: conifer-broadleaf, and beech (Wardle *et al.* 1983) (Fig. 2.8).

The conifer-broadleaf forests are predominantly lowland and are dominated by podocarps, including the genera *Dacrycarpus, Dacrydium, Phyllocladus* and, of course, *Podocarpus* (Wardle *et al.* 1983). Also present are two species of *Libocedrus* and the kauri *(Agathis australis).* The warmer inland forests support primarily *Podocarpus spicatus* and *Podocarpus totara* as canopy dominants, along with a more minor component of *Podocarpus dacrydiodes* and *Dacridium cupressinum* (Entrican & Holloway 1956). Hardwood associates of lower stature include the genera *Beilschmiedia, Knightia, Laurelia, Litsea* and *Nestegis* (Wardle *et al.* 1983). Forests range from relatively open to closed canopy, depending on species and habitat. For example, *Dacridium cupressinum, Podocarpus hallii* and *Podocarpus totara* are most common on drier sites and may form near-monospecific stands, while *Podocarpus dacrydiodes* often dominates on wetter, swampy sites (Cockayne 1958). In the coastal hills and plains, *Dacridium cupressinum* is commonly dominant, although *Podocarpus hallii* and *Podocarpus totara* are likely to occupy more xeric sites.

Genera common and often dominant in coastal lowland forests include *Elaeocarpus, Met-rosideros* and *Weinmannia* (Cockayne 1958; Wardle *et al.* 1983). The conifer-broadleaf shrub community in these forests is often well developed, with abundant species. The overstorey tree *Weinmannia racemosa* is abundant and widely distributed on the wetter, western side of both islands. Another important tree species is the giant kauri *(Agathis australis)*, much reduced in extent and growing primarily north of 38° S, particularly on ridges (Wardle *et al.* 1983). It grows with podocarps and various hardwoods (Cockayne 1958), and occurs both singly and in groups or thickets on poorer soils.

The beech, or *Nothofagus*, forests occupy cold, wet, montane environments and are characteristically lower in stature and less species-rich than the conifer-broadleaf forests. Four species of *Nothofagus* grow in New Zealand: *N. solandri, N. menziesii, N. fusca* and *N. trun-cata. Nothfagus solandri* is the most widespread of the four species and there are two recognized varieties: var. *solandri* and var. *cliffortoides*.

The former occupies warm, dry lowland environments primarily from East Cape on North Island to the centre of South Island (Cockayne 1958; Wardle *et al.* 1983), while var. *cliffortoides* grows at higher altitudes or on poor soils under high rainfall to the south. *Nothfagus solandri* forms nearly pure stands, with a sparse under-storey composed of shrubs (especially *Coprosma* spp.), lichens and mosses (Wardle *et al.* 1983), although both varieties co-occur where their geographical ranges overlap. *Nothofagus fusca* is distributed from East Cape to the southern end of South Island, and grows best on deeper, well-drained soils. *Nothofagus menziesii* has similar site requirements and dominates the subalpine zone of western South Island and the eastern mountain ranges of North Island (Wardle *et al.* 1983). *Nothofagus truncata* grows on drier, less fertile sites, primarily north of 42° S.

### *European*

Temperate forests once covered much of Europe and the British Isles, but most of the original forest has long since been converted to agriculture. Moreover, many introduced species have been extensively planted, for example plantations of the western North American conifers *Picea sitchensis* and *Pseudotsuga menziesii* in the British Isles. None the less, temperate forests remain an important resource and there are numerous recognized community types.

Jahn (1991) has extensively described the European deciduous forests, and I make considerable use of his excellent work in my more abbreviated summary here.

Two regions of temperate forest can be recognized: the Middle European, which occupies a band between about 43° and 55-60°N, and the Submediterranean, which generally follows the southern border of the Middle European south to a latitude of about 40° N.

*Middle European.* In the Middle European region are four somewhat distinctive provinces: the Atlantic, Subatlantic, Central European and Sarmatic. Throughout the region are genera common to most of the temperate forest of the Northern Hemisphere, including *Acer, Betula, Carpinus, Fraxinus, Pinus, Populus, Prunus, Quercus, Sorbus, Tilia* and *Ulmus*. The genus *Fagus* is represented principally by *Fagus sylvatica,* which is distributed throughout most of the Middle European region and is, alongside *Quercus,* the most conspicuous and perhaps most ecologically important tree in Europe.

The Atlantic province includes the British Isles and the coastal regions of western Europe, and a number of characteristic native species are, or were, widely distributed. In the British Isles, *Quercus* spp. still dominate many sites, although *Acer* spp., *Carpinus betulus, Fraxinus excelsior, Tilia* spp., and other species are abundant. *Fraxinus excelsior* is often the dominant species on high pH soils, with *Quercus robur* and *Quercus petraea* on sites of lower pH and fertility (Jahn 1991). *Fagus sylvatica* and *Fagus sylvatica-Quercus petraea* woodlands occur in south-eastern England. *Fagus sylvatica* is also common in north-western France, portions of Belgium and throughout the Middle European and portions of the Submediterranean regions. In the Atlantic province, *Fagus sylvatica* is typically absent only on sites that are wet, steep, calcareous or very sandy (Jahn 1991). The Subatlantic province is orientated along a north-east to south-west axis, covering an east-west region between about 8 and 15°W in southern Fennoscandinavia and about 3° W and 3°E in southern France and northern Spain. *Quercus robur* is common in the north and co-occurs with a number of associates depending on soil conditions (see Table 2.5). *Fagus sylvatica* is prominent on upland sites in the north and in the hills and mountains to the south, and forms a complex regional mosaic of communities, determined by a combination of climatic, soil and physiographic factors (Jahn 1991) (see Table 2.5). In the southern portion of the Subatlantic province, *Quercus* spp. become increasingly prominent, particularly at lower elevations. In the eastern Pyrenees, *Quercus pubescens* is found at the lowest elevations, *Fagus sylvatica* and *Abies alba* at higher elevations, with *Abies alba* increasing in importance with altitude (Jahn 1991).

The Central European province covers the former East Germany, western Poland and western Czech Republic (Jahn 1991) and is characterized by a more continental climate than the Atlantic or Subatlantic provinces. *Fagus sylvatica* reaches its easternmost extent here, and occurs extensively north of the Baltic Sea. It has many associates, depending on elevation, and often has a more species-rich ground layer than comparable forests in the Subatlantic province. *Carpinus betulus-Quercus petraea* communities formerly occupied rich, mesic plains and uplands, but most have been converted to agriculture. *Betula-Quercus* forests grow on nutrient-poor, low pH soils, with *Betula pubescens* and *Quercus robur* common on wet sites and *Betula pendula* and *Quercus petraea* on drier sites. *Pinus sylvestris* and *Picea abies* are locally intermixed in the north and east, with *Larix decidua* common to the hills of southern Poland (Jahn 1991).

The Sarmatic province extends in a band eastward to nearly 60° E, its north-south boundaries tapering from about 50-60° N in the west to about 53-54° N at its eastern edge. To the south lies steppe and to the north boreal forests (see p. 26). Generally, the Sarmatic province is characterized by a progressive west-to-east loss of the deciduous species characteristic of the three provinces to the west and an increase in the importance of conifers. For example, there is a tendency for nutrient-poor sites to shift from domination by *Betula* and *Quercus robur* in the Subatlantic province to domination by *Pinus* in the Sarmatic. Similarly, on more fertile sites, *Fagus sylvatica* is no longer present, and has been replaced by *Tilia, Quercus robur, Carpinus* and *Pinus. Pinus sylvestris* is distributed throughout the Sarmatic province and is an associate or dominant on all but the most fertile, moist soils. On sandy soils it forms nearly pure stands, without deciduous associates. *Picea abies* is also common, occurring with *Pinus sylvestris*, and increasing in importance with soil fertility and moisture. *Abies alba* occurs in the eastern part of the province, with *Abies sibirica* in the far western portion. *Populus tremula* and *Quercus robur* are present along the southern boundary, although the latter is uncommon in the north. In contrast to the northern portions of the Sarmatic province, the south is largely deciduous. *Quercus robur* and *Tilia cordata* are dominant, but *Carpinus betulus* is an important (though often low stature) associate.

Intermixed or adjacent to these four provinces of the Middle European region are the forests of the Alps and the Carpathian Mountains. The Western Alps (orientated north-south and running from the Mediterranean to Grenoble, France) are occupied by *Quercus pubescens* forests in the west-facing foothills, with *Fagus sylvatica*, *Quercus petraea* and *Pinus sylvestris* occurring at higher elevations. High-elevation interior forests contain a mix of deciduous and coniferous species. East-facing forests are predominantly deciduous hardwoods, with a large *Quercus* component.

In the Middle Alps, the north-facing foothills are forested with *Quercus petraea* and *Carpinus betulus*, which are replaced by *Fagus sylvatica* and *Acer pseudoplatanus* at higher elevations, *Abies alba* and *Picea abies* in the interior, *Fagus sylvatica* on upper south-facing slopes and *Cas-tanea sativa* and *Quercus petraea* on the south-facing foothills. The Eastern Alps (Verona, Italy to the Dinara mountains of the former Yugoslavia) are occupied by a variety of hardwoods, with an increasing abundance of conifers at higher elevations. Similar species occur on the south-facing slopes, with the addition of *Fraxinus ornus*, *Ostrya carpinifolia* and *Quercus* spp. in the low foothills and foothill-plains transition.

The Carpathian Mountains extend in an arc from Slovakia east through south-west Ukraine and into central Romania. The forests of the foothills and submontane zone (up to about 600 m) of the western Carpathians are predominantly mixed hardwood forests, with a minor component of *Fagus sylvatica*. On the eastern side, forest and steppe alternate at the lowest elevations, and are progressively replaced by mixed hardwoods, and then coniferous forests, as elevation increases (Jahn 1991).

*Submediterranean.* The Submediterranean region is characterized by a preponderance of deciduous oaks (*Quercus* spp.) and represents the southern distribution of many of the species common to the mixed forests of the Middle European region to the north. To the south, where summer droughts are more extended, grow the evergreen Mediterranean forests. The vegetation is quite variable and complex throughout and is sensitive to climatic and site factors in this region, which borders temperate and Mediterranean climatic zones. Deciduous species of the Submediterranean region include *Carpinus orien-talis, Fagus sylvatica, Fagus moesica, Fagus orientalis, Fraxinus ornus, Ostrya carpinifolia, Quercus cerris, Quercus faginea, Quercus frainetto, Quercus pubescens* and *Quercus pyren-aica.* Various species of *Acer* and *Sorbus* are also present. *Fagus* spp. are characteristic of montane regions and co-occur with *Abies* spp. and *Juni-perus* spp. (Jahn 1991). *Pinus* spp. are present, with *Pinus nigra* growing with *Quercus cerris, Quercus faginea, Quercus pubescens* and *Quercus pyrenaica* in submontane environments, *Pinus sylvestris* common in montane zones, and *Pinus peuce, Pinus heldreichii* and *Pinus mugo* growing at high elevations.

### *Near Eastern*

The Near Eastern temperate forest lies between 35 and 45° N, from about 27° E in eastern Turkey to about 58° E in eastern Iran. The forests of this region are a mixture of both European and Colchian species (Röhrig 1991b), with elements of (sub)Mediterranean and steppe flora.

In western Turkey, oak forests are common, composed of various species of *Quercus* and other hardwoods. *Fagus orientalis* grows at higher elevations, though not in abundance. In the western portion of the Anatolian plateau in Turkey, *Quercus*-dominated temperate broadleaf forests grow between 1000 and 2000 m (Röhrig 1991b) and contain a number of other genera as well. Extending into Iran further to the east are other forests dominated by *Quercus* (Röhrig 1991b). Various species of *Acer, Pyrus* and *Prunus* also occur, as well as a few species, like *Fraxinus excelsior,* more commonly found in the Middle European and Submediterranean regions. On alluvial soils at lower elevations (<400m) grow more diverse forests, which in addition to *Quercus* spp. include *Alnus barbata* and *Carpinus* spp. Moist forests between the Caspian Sea and Black Sea contain a number of species, including many missing from the European forests to the west (Walter 1985). Further east, the species and assemblages change but a few genera predominate, with a number of low-stature woody species occupying the understorey (e.g. *Buxus, Crataegus, Corylus, Ilex, Ligustrum, Rhododendron* and *Sambucus)* (Röhrig 1991b).

At high elevations (600-1100 m) are the beech *(Fagus orientalis)* forests. Near pure *Fagus orientalis* forests grow between the Sakarya River in western Turkey and the Turkey-Georgia border. Between 600 and 1300 m on calcareous northern slopes on the mountains of the southern Crimean Peninsula, *Fagus orientalis* grows alone or with *Carpinus betulus, Fraxinus excelsior, Tilia cordata* and *Ulmus glabra* (Röhrig 1991b). On southern slopes, it is found with *Acer hycranum, Pinus nigra, Carpinus orientalis* and *Quercus petraea.* In the Caucasus Mountains, it grows between 1000 and 1500m with *Acer* spp., *Fraxinus excelsior, Tilia platyphyllos* and *Ulmus glabra.* At higher elevations, *Fagus orientalis* grows with coniferous species like *Abies born-mülleriana, Abies equitrojani, Abies nordmanni-ana, Picea orientalis* and *Pinus sylvestris,* along with some of the broadleaf species found at the lower elevations.

### *North and Central American*

Temperate forests occupy much of North America, from north of the border between the USA and Canada south into Mexico and some Central American countries at high altitudes. The conifer-dominated forests of western North America are separated from the more deciduous forests of the east by vast areas of former prairie now dominated by agriculture where rainfall or irrigation permit.

*USA and Canada.* The eastern North American temperate forest is primarily deciduous, although the forests of the south-eastern USA contain large amounts of conifers, both natural and planted, as do the states bordering the Great Lakes. The forests of the region have been extensively described, especially by Braun (1950) and more recently Barnes (1991). I follow Barnes's (1991) classification scheme, which is derived from Braun's (1950) and Küchler's (1964). The forest regions described below are roughly confined to specific geographical areas, to which I refer in each section in an attempt to maintain some consistency with the other sections in this chapter.

The South-eastern Evergreen region covers the coastal plain of the south-eastern USA, stretching roughly from Virginia along the Atlantic coast, continuing south across Florida and the Gulf of Mexico to Texas. *Pinus palustris-Aristida stricta* savannahs formerly dominated much of the uplands but have been largely converted to other uses or cover types, and both *Pinus elliottii* and *Pinus taeda* are presently the most abundant conifers in the region. *Magnolia grandiflora* and *Fagus grandifolia* are believed to be climax dominants on mesic sites, although a number of other genera are abundantly represented.

Inland of the South-eastern Evergreen region is the Oak-Hickory-Pine region, extending from Pennsylvania to Alabama and bordered by the coastal plain towards the Atlantic Ocean and the Appalachian mountains to the west. As its name implies, the principal species include those in the genera *Quercus* and *Carya*, often in mixture with species common to adjacent regions, including *Pinus*.

The Appalachian Oak region includes the Appalachian mountains, from Georgia north and eastward into eastern New York and Massachusetts, and is the most floristically diverse in North America. Oaks are abundant at all elevations and aspects. At higher elevations (> 1100 m), species are more characteristic of the beech-sugar maple and sugar maple-basswood forests to the north (see below). At the highest elevations are boreal forests composed of *Picea rubens* and the endemic *Abies fraserii* in the south, and *Picea rubens* and *Abies balsamea* to east to south-west from western Pennsylvania to northern Alabama, and coincides physiographi-cally with much of the unglaciated Appalachian plateaus. The vegetation comprises principally species from the regions which it borders. The Western Mixed Mesophytic region lies immediately to its west, covering central and western Tennessee and Kentucky, the southern boundary of the Wisconsinan Glaciation in southern Indiana and Ohio, and northern Mississippi and Alabama. Forests dominated by *Quercus* spp. cover most of the region, with a large component of *Carya* and *Pinus*.

The westernmost region is the Oak-Hickory, bordered to the west formerly by prairies and presently by agricultural lands. It extends from Manitoba, Canada to southern Texas. South of the glaciated portion lie the Ouachita Mountains and Ozark Plateau. *Quercus alba* is the dominant species on dry-mesic uplands, with various species of *Carya*, *Quercus* and *Pinus* characteristic of drier sites. In the glaciated region to the north, *Quercus* spp. are common on drier uplands, while *Fagus grandifolia* and *Acer sac-charum* grow on more mesic sites. *Quercus macrocarpa* formerly formed savannahs interspersed with prairie, most of which has been lost to agriculture and urbanization.

The Sugar Maple-Beech region occurs on glaciated till in the southern half of Michigan's lower peninsula, Indiana and Ohio; *Acer saccha-rum* and *Fagus grandifolia* are the dominant species in mature forests, although a number of other species are common (see Table 2.8). The Sugar Maple-Basswood region to the north of the Sugar Maple-Beech region is situated on glaciated terrain and, as the name implies, *Acer saccharum* and *Tilia americana* are dominant. Other associates are those characteristic of the Sugar Maple-Beech region (see Table 2.8).

The Northern Hardwood-Conifer region extends across the northern USA and southern Canada from Manitoba to Maine and Nova Scotia. The terrain is glaciated and supports a wide array of forest types. Species common to the Sugar Maple-Beech region predominate on glacial till, although the region extends beyond the northern range of *Carya*, *Liriodendron*, *Nyssa* and other genera common to the more southern forest regions. *Quercus ellipsoidalis* grows on dry glacial outwash plains, but fire-prone *Pinus banksiana* is the most dominant species on these soils. Much of the presettlement forest was dominated by *Pinus strobus* but virtually all of the original forest was cut and converted naturally to hardwoods, including *Betula papyrifera*, *Populus tremuloides*, *Populus grandidentata* and *Quercus* spp. Coniferous swamps on calcareous soils often consist predominantly of *Thuja occidentalis* with lesser amounts of other hardwoods.

The temperate forests of western North America differ from those in the east by a preponderance of conifers rather than hardwoods. Furthermore, much of forested western North America is mountainous and the vegetation is often distributed in rather discrete elevational bands, as opposed to the more subtle changes with landscape position associated with the eastern deciduous forests.

The Rocky Mountains extend from north-central Canada, through the USA and into Mexico, and form the eastern border of North America's temperate coniferous forests. At the lowest elevations (1500-1600m), *Pinus edulis* and *Pinus monophylla* grow with various species of *Juniperus* and *Quercus*. At higher elevations are *Pinus ponderosa* forests, both closed canopy and open park-like stands, above which lie forests dominated by *Pseudotsuga menziesii*. In the southern Rockies, the associates of *Pseudotsuga menziesii* are few, but in the northern Rockies it is variously found with *Abies grandis*, *Larix occidentalis*, *Pinus contorta* and *Populus tremuloides*. With increasing elevation in the northern Rockies, *Pseudotsuga menziesii* occurs with *Thuja plicata* and *Tsuga hetero-phylla*. These two species often form nearly pure stands. *Populus tremuloides* grows at similar elevations, colonizing after disturbances, principally fire. Just below the treeline are found *Picea engelmannii* and *Abies lasiocarpa*. In the far northern Canadian Rockies, the principal sub-alpine species are *Abies lasiocarpa*, *Picea glauca* and *Pinus contorta*.

In the Sierra Nevada Mountains of California, *Pinus sabiana* is a dominant conifer in the western foothills, growing with *Quercus* spp. On western slopes, *Pinus lambertiana* grows with *Pinus ponderosa* at higher elevations (750-1800 m), together with *Abies concolor* and *Calo-cedrus decurrens*.

*Pinus jeffreyi* occupies lower eastern slopes and higher west-facing elevations. Further south, *Pseudotsuga macrocarpa* and *Pinus coulteri* grow in the elevations between desert scrub and the mixed conifer forests described above. *Sequoiadendron giganteum*, the world's largest trees, are distributed between 1350 and 2250 m on the western slopes of the Sierra Nevada. *Abies magnifica* and *Pinus contorta* grow between roughly 1800 and 2400 m, in regions of extremely high snowfall (4-20 m snow-pack). The subalpine forests (>2400m) consist of *Pinus albicaulis*, *Tsuga mertensiana*, *Pinus balfouriana*, *Pinus flexilis* and the long-lived (>4000 years) *Pinus aristata*.

The coastal mountain ranges of the northwestern USA and south-west Canada differ from the interior mountains principally by the abundant precipitation they receive (upwards of 2000 mm annually) and the exceptionally large size that many of the species attain. Along the Pacific coast from northern California into southeastern Alaska and the Kodiak Islands, *Picea sitchensis* forests are found in narrow bands (generally only a few kilometres wide) below 150 to 600 m elevation (Franklin & Dyrness 1973). Its associates include *Pseudotsuga menziesii, Thuja plicata, Tsuga heterophylla* and *Alnus rubra.* In southern Oregon and northern California, *Sequoia sempervirens* forests grow at similar elevations, extending as far as 15-20 km inland, and growing on slopes and sometimes at higher elevations.

Above the *Picea sitchensis* zone is the *Tsuga heterophylla* zone (550-1200m) (Franklin & Dyrness 1973), above which *Abies amabilis* dominates the elevational bands ranging between 600 and 1500 m. The highest elevation forests are the subalpine *Tsuga mertensiana* forests that grow as high as 2000 m in southern Oregon. Principal associates include *Abies amabilis, Abies lasio-carpa* and *Abies procera. Tsuga mertensiana* is replaced by *Abies lasiocarpa* in the interior portions of the Cascades and on the drier eastern slopes. The most extensive forests of the eastern slopes of the Cascades are those dominated by *Abies grandis,* which grow at elevations of 1000-1500m in the north and 1000-2000m in southern Oregon.

*Central America and Mexico.* Geographically speaking, the forests of the region are primarily (sub)tropical in nature, although there is a large degree of taxonomic similarity between the vegetation of the USA and Mexico and Central America. Some of the similarity is due to the continuity of species' range distributions, but it is also believed that the region provided a refuge during cool periods of the Oligocene, after which geological uplift and further climatic changes isolated the temperate forests to disjunct islands among otherwise (sub)tropical forests from Mexico southward into Nicaragua (Röhrig 1991d).

Among deciduous species, *Carpinus carolini-ana, Liquidambar styracifl.ua, Nyssa sylvatica, Ostrya virginiana* and *Prunus serotina* grow in both the USA and Mexico. Species with a high degree of similarity include *Fagus mexicana* and *Fagus grandifolia, Carya ovata* var. *mexicana* and *Carya ovata*, and *Acer sutchii* and *Acer saccha-rum*. Several genera are well represented in both regions, including *Alnus, Carya, Fraxinus, Juglans, Magnolia, Platanus, Populus* and *Salix*. The genus *Quercus* is especially abundant, both deciduous and evergreen.

In montane deciduous forests, *Juglans olancha, Liquidambar styraciflua* and *Ostrya virginiana* are dominant, growing at 1200-1800m in the Sierra de Chucharas in the Mexican state of Tamaulipas. Forests dominated by *Quercus* spp. and *Pinus* spp. often occupy the highest elevations (> 1800 m), with (sub)tropical species prominent in the lower elevations (Röhrig 1991d). Riverine forests may be composed primarily of deciduous genera like *Carpinus, Fraxinus, Platanus, Populus, Salix* and *Ulmus*, but uplands typically contain representatives from more (sub)tropical families like Clethraceae, Fabaceae, Sabiaceae, Lauraceae, Staphylaceae, Cunoniaceae and Rutaceae.

Mexico and Central America also contain a rich coniferous flora, especially of pines. There are at least 47 *Pinus* species (Perry *et al.* 1998), which is about half the world total. Many species are five-needled pines, and they occupy niches from coastal to high montane environments. Some, e.g. *Pinus caribaea* and *P. patula*, are widely planted in the tropics.

### *South American*

The South American temperate forests are the smallest in extent, confined primarily to Chile within and between the coastal and Andean mountain ranges, with a small extension into Argentina. Both broadleaf evergreen and deciduous species are present, often in mixture, although evergreen species are dominant. *Nothofagus* spp. are probably the most abundant trees; a total of 10 species grow in the region, of which seven are deciduous (*N. obliqua, N. alpina, N. pumilio, N. antarctica, N. glauca, N. leoni* and *N. alessandri*) and three evergreen (*N. betu-loides, N. dombeyi* and*N. nitida*). The only other deciduous species present is *Acacia cavens*, which grows in a more Mediterranean climate in the northern part of the temperate region, between about 31 and 38° S.

*Nothofagus glauca* and *N. obliqua* grow in the wetter climate to the south of the *Acacia* forests in central Chile, the former on eastern and northern slopes and the latter on lowland sites. *Nothofagus alessandri* mixes with *N. glauca* on moister sites of the coastal mountains (Schmaltz 1991). *Nothofagus obliqua* dominates much of the lowlands and valleys south of 36-38° S, depending on proximity from the ocean. It is often found with *N. alpina* and *N. dombeyi*, the latter of which often forms pure stands in cool wet gorges (Schmaltz 1991).

The conifer *Libocedrus chilensis* is often present on dry slopes opposite *N. obliqua*, and *Laurelia sempervirens* and *Persea lingue* are common, though lower-stature, associates of *N. obliqua*. At higher elevations in the Andes, *N. dombeyi* replaces *N. obliqua*. *Nothofagus alpina* forests occur in narrow elevational bands in the northern Chilean Andes and the species is widespread in mixed forests of the coastal range, often found with *N. dombeyi* and the associates of *N. obliqua* (Schmaltz 1991). *Auracaria araucana* occurs over small areas of the coastal range and is more widespread in the Andes (37-41° S) above the *N. alpina* forests (600-900m). *Nothofagus pumilo* occupies cold snowy sites, at elevations of 1500-1800 m at about 37° S down to sea level at Cape Horn. *Nothofagus antarctica* is a high-elevation krummholz-forming species that also colonizes primary successional habitats (e.g. volcanic deposits) and other harsh or exposed sites (Schmaltz 1991). *Nothofagus nitida, Podocarpus nubigenus* and *Weinmannia trichosperma* occupy humid wetter sites between 600 and 900 m elevation south of about 40° S. *Fitzroya cupres-soides* is a large-sized tree occurring on wetter sites between about 40° and 43° 30'S in both the Andean and coastal ranges, and at elevations from about 700m to the timberline (Veblen *et al.* 1983).

## SAVANNAHS AND OPEN WOODLANDS

Savannah is a broad term used to describe a number of ecosystem types. Generally speaking, savannahs are ecosystems dominated by a grass ground cover with an overstorey of widely spaced trees that do not form a closed canopy. Cole (1986) described five types of savannah.

1. Savannah woodlands have a canopy of large-stature deciduous or semideciduous trees (>8m) with tall grasses (>80cm) and grow in mesic environments.

2 Savannah parklands also consist of tall grasses but with scattered low-stature trees (<8 m).

3 Savannah grasslands lack tree or shrub cover.

4 Low tree/shrub savannahs contain widely spaced low-stature grasses (<80cm) and trees or shrubs (<2m).

5 Thicket/scrub savannahs consist of medium-stature trees that lack a grass understorey.

Savannahs have been substantially altered by humans on a global scale. Conversion to agriculture, especially livestock grazing, has dramatically altered community composition and structure. These problems have been most acute in less-developed nations, where the climate is often the most arid and where heavy demands for wood for fuel and shelter, combined with grazing, have led to severe resource overexploitation. However, even the North American savannahs have been subjected to heavy use. For example, the *Pinus palustris* forests of the south-eastern USA now occupy less than 1% of their original habitat, the rest having been converted to agriculture, urban environments or plantations of other species.

Fire, caused by both lightning and humans, has long been an important force in shaping the structure and composition of savannah ecosystems. Generally, fire promotes the regeneration of the grass layer while inhibiting the establishment of trees. Humans have deliberately set fire to Africa's savannahs for at least 50000 years (Rose Innes 1972) and for perhaps 40000 years in Australia (Nicholson 1981), and this has undoubtedly promoted their persistence and expansion. Fires occur naturally during the dry season, but most fires in savannahs are kindled intentionally in order to clear land for ploughing or promote palatable new growth for livestock. Fire interacts with climate and soils, though, and is often regarded as a secondary causative factor in the establishment of savannahs. For example, humans have been burning savannahs for only about 5000 years in South America, an insufficient length of time to account for their broad expanse there. However, there is no question about the importance of fire in maintaining the open structure and grass understorey of pre-existing savannahs. In North America, for example, the large expanse of prairie and savannah (dominated by *Quercus macro-*

*carpa)* that extended into the midwestern USA at the time of the European settlement is generally ascribed to the action of fire, particularly those set by humans. These areas subsequently reverted to forest after European settlement and fire suppression, and few remain today.

Savannahs are extensively grazed throughout much of the world. Prior to their exposure to domesticated livestock, the savannahs of Africa in particular had been subjected to extensive pressure from wildlife grazing for much of their 25 million year history. Grazing generally inhibits the regeneration of woody plants, but where there is a long history of intensive selective herbivore pressure on plants, such as Africa, many woody species have evolved both chemical and morphological defences against grazers and browsers. Intensive grazing by introduced domesticated livestock with which the plants have not evolved has led to the replacement of native species by exotics in some regions. As is the case with fire, though, grazing alone is insufficient to explain the presence and maintenance of savannahs.

## Distribution and extent

Savannahs occupy about 25 million km$^2$, most of which (24 million km$^2$) are found in the tropics. Although there are about 700000km$^2$ of savannah in North America, I focus my discussion here on the tropical regions. I have described some of the most important temperate savannahs, like the *Pinus palustris* forests of the southeastern USA and the pinyon-juniper woodlands of the western USA, in the section on temperate forests. McPherson (1997) has extensively described the North American savannahs and I refer the interested reader to his text.

Africa contains most of the world's savannahs, about 15 million km$^2$ (Johnson & Tothill 1985), most of which occupies a broad latitudinal area between about 29° S and 16°N. A band of savannah ecosystems growing between the equator and 20° N covers, or once covered, the continent nearly from coast to coast. The miombo woodlands of eastern Africa consist of about 3 million km$^2$ and constitute the largest single vegetation type in the region. In Australia, savannahs occur between 11 and 30° S, stretching across much of northern Australia and curving south into southeastern

Queensland. The Asian formation formerly covered much of eastern India, although it has largely been cleared for agriculture. In South-East Asia, savannah is interspersed with tropical forest, and is found in parts of Cambodia, Laos and Vietnam and to a lesser extent in Thailand and Burma (Archibold 1995).

The South American savannahs occur between 25° S and 23°N, with the largest portion composed of the Brazilian cerrado (1.8 million km$^2$), the pantanal (400000km$^2$) in south-west Brazil adjacent to Bolivia and Paraguay, the chaco (800 000 km$^2$) of western Paraguay, eastern Bolivia and northern Argentina, and the Llanos de Orinoco in western Colombia and central Venezuela (Medina & Silva 1991; Solbrig 1996). Lesser areas include the Llanos de Orinoco Moxos of Bolivia and the Andean foothills, and the Gran Sabana of Venezuela.

## Climate

Savannahs are characterized by a largely tropical temperature regime and an extended dry season, typically in winter. Mean annual temperatures range from approximately 19 to 27°C, and reach their maximum at the end of the wet season (Archibold 1995). Mean warm season temperatures can exceed 30°C. Annual precipitation ranges from 500 to 1500 mm, although the dry season may last for 6-7 mcnths, during which only a small percentage (<5%) of annual precipitation may fall. The rainy season typically lasts from late spring through autumn.

## Soils

Savannah soils in the higher-rainfall regions are often Oxisols: old, highly weathered, high in clay and somewhat acidic (pH 4.8-5.2). In dryer climates, Alfisols are more common, as found in North and East Africa, India and portions of Brazil. Entisols are found in the driest savannahs, such as those in southern Africa and northern Australia. They are characteristically low in phosphorus and often gravelly or stony. Vertisols are found in parts of eastern Australia. They tend to be high in montmorillonite clay, low in phosphorus, somewhat saline and have a low pH.

Many savannah soils have carbon concentrations in the range 1.5-3.0% (Lopes & Cox 1977); savannah soils as a whole contain about 14% of the world's soil carbon (Kirchmann & Eklund 1994). As stated above, however, nutrient contents are often low, and internal cycling is efficient and quantitatively important. Efficient cycling of critical nutrients like nitrogen, combined with atmospheric inputs, compensate for losses due to fire and leaching (Kellman 1989).

## Regional floras

### *African*

Between moist tropical forests and the shrub-steppe ecosystems that border deserts are a wide variety of dry forests, including both open woodlands and savannahs. The miombo woodlands of southern, eastern and central Africa are dominated by the legumes *Brachystegia* spp. and *Isoberlinia* spp. (Aubrévillé 1956). Various species of *Baikiea, Cryptosepalum* and *Burkea* vary locally in their importance. Representatives of the Euphorbiaceae *(Uapaca)* and Dipterocarpaceae *(Monotes* and *Marquesia)* are of minor importance. Forests of *Baikiea plurijuga* grow on Kalahari sand formations in portions of Zimbabwe. The forests of south-western Africa are less extensive and less species-rich, the better forests being dominated by *Isoberlinia* and *Uapaca*. To the north in West Africa, *Anogeissus leiocarpus* and *Lophira lanceolata* grow with an understorey of *Andropogon, Hyparrhenia* and *Pennisetum* spp. (Archibold 1995). Forests of *Adansonia digitata* and *Sclerocarya birrea* grow to the east. In more arid regions of southern Africa, *Colophospermum mopane* forms nearly pure stands, and shrubby *Acacia* savannahs occur on drier soils as well. Palm savannahs occur on more poorly drained sites, principally *Borassus* spp. (Africa) and *Medemia* (Madagascar).

### *Australian*

*Eucalyptus* and *Acacia* spp. dominate most of Australia's savannahs and open woodlands. In northern Australia, *Eucalyptus tetrodonta* is widespread where annual rainfall is 700-1500mm (Beadle 1981). The graminoids *Aristida* spp., *Chrysopogon latifolius, Heteropogon trit-iceus, Sorghum* spp. and *Themeda australis* are common in the understorey. Other *Eucalyptus* species occurring in high-rainfall areas include *E. miniata* and *E. polycarpa,* which grow with *E. tetrodonta, E. tectifica, E. confertiflora* and *E. grandiflora* on finer soils, and with *E. ptychocarpa* in riparian environments. In the drier areas to the south, *E. argillacea* and *E. terminalis* grow in fine-textured soils and *E. dichromophloia* occurs on stony soil and deep sands. *Eucalyptus crebra* and *E. drepanophylla* are common in north-eastern Queensland. Low-stature *Melaleuca viridiflora* and *Melaleuca nervosa* woodlands occur in the north and in eastern Queensland (Gillison 1994), along with a number of *Eucalyptus* spp. *(E. brevi-folia, E. dichromophloia, E. jensenii, E. papauana* and*£. prunosa).*

Further inland, eucalypts are replaced by low-stature *Acacia* shrubland. However, *Acacia* spp. also form open woodlands in areas of higher precipitation. In central Queensland, *Acacia harpophylla* woodlands (brigalow) reach nearly to the coast, with *Brachychiton rupestris* a common associate. In north-eastern Australia, *Acacia* woodlands and savannahs grow on both coarse, shallow, acid soils and on deep, fertile, alkaline soils (Johnson & Burrows 1994). *Acacia shirleyi* and *Acacia catenulata* often dominate the poor soils, with *Acacia harpophylla* and *Acacia cam-bagei* common on the better sites. *Acacia aneura* grows along the coast of Western Australia with short tussock grasses, especially *Eragrostis erio-poda* and *Monochather pardoxa.*

### *South American*

South America's largest region of savannah is the cerrado of east-central Brazil, occurring predominantly on tablelands at elevations of 300-1000m. Dominant grasses include *Andro-pogon, Aristida, Paspalum* and *Trachypogon* (Archibold 1995). Associated tree species include *Caryocar brasiliense* and *Curatella americana.* In the transition zone between closed forests and savannah, species such as *Dipteryx alata, Dilodendron bipinnatum, Luehea paniculata, Magonia pubescens* and *Pseudobombax tomen-tosum* are characteristic (Ratter 1992).

The llanos occur at lower elevations in the floodplain of the Orinoco River in Venezuela and Colombia. Grasses and sedges of the genera *Panicum, Leersia, Eleocharis, Luziola* and *Hymenachne* dominate, with trees widely scattered and often forming small 'islands' (Blydenstein 1968; Walter 1985). Tree species include

*Bowdichia virgilioides, Byrsonima cras-sifolia* and *Curatella americana.* The llanos have a pronounced seasonal wet-dry climate, often flooded during the wet season, while drought of 5 months or more is not uncommon during the dry season. The swampy pantanal lies to the south of the llanos and has many of the same species. Common grasses include *Panicum, Paspalum* and *Sorghastrum* spp., and the tree *Curatella americana* is common on wetter sites.

In Central America, the most abundant savannah species is probably *Pinus caribaea,* although several species of *Quercus* also occur. *Pinus caribaea* savannahs dominate much of southern Belize, with associated palms, cecropias and miconias (Kricher 1997).

### *Indian and Asian*

Much of eastern India was formerly savannah but little remains; most has been cleared for agriculture. The small trees *Anogeissus latifolia* and *Diospyros melanoxylon* occur in small woodlands, and thorny species like *Acacia catechu* and *Zizyphus jujuba* have increased in response to heavy grazing pressure (Archibold 1995). The remaining savannahs of South-East Asia are dominated by deciduous dipterocarps. Savannahs typically occur where the dry season is 5-7 months in length and precipitation varies between 1000 and 1500 mm (Rundel & Boonpragob 1995), predominantly in hilly or mountainous terrain. Dominant species include *Shorea obtusa, Shorea siamensis, Dipterocarpus obtusifolius* and *Dipterocarpus tuberculatus.*

## MEDITERRANEAN ECOSYSTEMS

The vegetation of Mediterranean ecosystems is characteristically low in stature, with a relatively open canopy. Evergreen shrubs and sclerophyl-lous trees are the dominant plant life forms, and frequent fires have historically played an important role in regulating community composition and structure. Most of these ecosystems have been heavily disturbed, subject to agricultural and horticultural conversion, grazing, cutting for wood and charcoal, and a long history of urbanization.

### Distribution and extent

Mediterranean ecosystems occupy about 1.8 million km$^2$, distributed among five geographical regions. In North America,

Mediterranean ecosystems occur along the southern coast of California and portions of northern Baja California, Mexico, between roughly 30 and 40° N. In South America, they occur along the coast of Chile between 30 and 39° S. Much of the coastal region surrounding the Mediterranean Sea in Europe, the Middle East and Africa is occupied by the ecosystem type bearing its name. The northern coast, from Portugal to Turkey, is (or was) predominantly Mediterranean woodland, as are portions of Libya, Morocco, Algeria and Tunisia. Southern African Mediterranean forests include both the coastal and mountain fynbos of the Cape in South Africa. The remaining Mediterranean ecosystems occur in south-western Australia, and portions of South Australia and Victoria.

## Climate

The Mediterranean climate is characterized by summer drought and wet cool winters. Annual precipitation ranges between about 300 and 1250 mm, of which two-thirds or more falls during the winter. Summer temperatures are warm, with July mean values of 20-30°C, and daily temperatures above 35°C are not uncommon. Winters average 10-12°C (Archibold 1995) and frosts are infrequent in all but the higher elevations. Relative humidity is often low, and in combination with strong seasonal winds and the sclerophyllous vegetation promotes hot, fast-moving fires.

## Soils

Mediterranean ecosystems are characterized by heterogeneous topography, including mountains, steep-sided canyons and intervening valleys, plains and tablelands. Consequently, the soils are highly variable. Generally speaking, soils are thin and low in available nitrogen and phosphorus, with relatively little horizon differentiation. Surface horizons tend to be neutral, with higher pH values in deeper horizons. A proposed sequence of decreasing fertility has been proposed: Chile > Mediterranean Basin > California > South Africa > Australia. Although based on data from macronutrient sampling from a few locations, it is reasonably accurate, at least with respect to Chile, California and Australia (Lamont 1995). However, broad geographical generalities about particular regions being characteristically low in a particular nutrient (e.g. phosphorus in Australia) are not necessarily especially widely applicable.

## Regional floras

### *Australian*

Mediterranean *Eucalyptus* forests occur in southwestern Western Australia, coastal South Australia and parts of Victoria. In Western Australia, Beadle (1981) recognized 14 alliances in the region, generally distributed as follows. In regions with relatively high rainfall (1000-1500 mm), *Eucalyptus diversicolor* and *E. jacksonii* dominate on fertile granitic soils, with *E. marginata* and *E. calophylla* on laterite soils and *E. gompho-cephala* on calcareous soils of the coast. In drier areas, *E. wandoo* and *E. loxophleba* replace these species, with *E. dundasii, E. lesouefh, E. salmonphloia* and *E. slubris* on the most xeric soils. *Eucalyptus cornuta* and *E. occidentalis* occupy wet soils. The mallee, a region dominated by eucalypts of shrubby morphology, forms the Mediterranean forests of South Australia and Victoria. There are about 100 species of *Eucalyptus* in the region. On deeper soils are found species like *E. diversifolia, E. dumosa, E. foecunda, E. gracilis, E. incrassata, E. loxophleba, E. oleosa, E. socialis* and *E. viridis.*

### *Chilean*

Mediterranean forests are confined to coastal Chile, where they occur from the lowlands up to about 2000 m elevation, at which point they merge into alpine communities. The flora is diverse, with nearly 2400 identified species. The herbaceous and shrub communities share a large degree of taxonomic affinity with Californian species, but the trees are relatively unique. *Salix, Prosopis* and *Rhamnus* are found in both regions, while *Gormotega, Jubaea, Legrandia* and *Pitvia* are endemic to Chile (Arroyo *et al.* 1995). The Chilean matorral comprises primarily low-stature (1–3m) evergreen species and occurs on the interior slopes of the coastal mountains and the Andean slopes north of 36° S and along the coast south of 32°S (Arroyo *et al.* 1995). Dominants include *Colliguaja* spp., *Cryptocarya alba, Lithrea caustica, Peumus boldus* and *Quillaja saponaria.* On moister sites *Salix chilensis* and *Cryptocarya alba* assume greater stature, with a shrub understorey. In Andean valleys the matorral grades into tall dry forests, and the two share many of the same species. To the north, the shrubs *Bahjia ambrosioides*, *Fuschia lycioides* and *Adesmia* spp. appear, and the vegetation is more scrubby and open. Scattered *Acacia covens* grow above herbs and grasses in the interior espinal (Archibold 1995).

### *Mediterranean Basin (European and North African)*

Because of the long history of human occupation and the extensive nature of human disturbance, much of the natural Mediterranean forest in the region has been lost. The remaining ecosystems share many of the species that occur in the Submediterranean region of Europe (see p. 46) and grade into these forests and temperate forests as elevation and/or distance from the coast increase.

In Spain and Portugal, *Quercus ilex* and *Quercus suber* are the dominant hardwoods, the latter occurring on non-calcareous soils. In general, these forests are heavily disturbed and rather sparsely populated with trees. The pines *Pinus halepensis*, *Pinus pinea* and *Pinus pinaster* grow at higher elevations. *Quercus ilex* is the most abundant species in the region, accounting for over 20% of forestland (Giordano 1956).

The mix of species in Italy is similar, and *Quercus ilex* is often found in mixture with *Arbutus unedo*, *Erica* spp., *Phylleria variabilis*, *Pistacia lentiscus* and *Myrtus communis*. *Pinus pinea* grows on sandy coastal sites, and with *Pinus halepensis* and *Pinus pinaster* inland. In Greece, *Quercus ilex* is common, being replaced by *Quercus coccifera* on calcareous soils. The mix of associates is similar to the rest of the region, with the addition of *Arbutus andrachne* and *Olea europaea* among others (Giordano 1956). Conifers in the region include *Pinus halepensis*, *Pinus brutia* and *Cupressus sempervirens*. In Turkey, *Quercus macrolepis* and other *Quercus* spp. occur, along with *Fagus orientalis* and the conifers *Pinus brutia* and *Pinus nigra* at lower elevations and *Cedrus libani* and *Abies cilicica* at higher elevations. In the coastal regions of the Middle East, *Quercus calliprinos* is common, along with many of the oaks and pines characteristic of forests to the west. Along the north coast of Africa, *Quercus ilex* is again often the dominant species. Associates include *Pinus halepensis* and *Abies numidica* at lower elevations and *Cedrus atlantica* above 1200m (Archibold 1995).

*North American*

Mediterranean ecosystems occur primarily in the coastal region of California and at low elevations in the Sierra Nevada. The coastal and desert-margin arborescent or shrubby vegetation (chaparral) covers about 3.4 million ha, about 20% of which is dominated by small trees. The shrub *Adenostoma fasciculatum* is common between 300 and 1500 m, and *Artemisia californica*-dominated coastal sage scrublands occur at lower elevations. *Ceonothus cuneatus* occurs with *Adenostoma fasciculatum*, and *Arctostaphyllos* spp. are common at higher elevations (Archibold 1995). On more mesic sites, trees are principally low-stature *Quercus* spp., including *Q. dumasa, Q. agrifolia* and *Q. wilslizenii*. There are about 3 million ha of open *Quercus*-dominated woodlands in California, not all of which experience a strong Mediterranean climate, but all of which have an open structure, are fire prone and are affected by increased demand for grazing land and urban expansion.

*South African*

The Mediterranean ecosystems in South Africa are represented primarily by the fynbos. At lower elevations (300-700 m) and on drier sites, these are primarily mid-stature (>2m) shrub communities, dominated by the genus *Protea. Protea neri-ifolia* is common, as are *Protea repens, Protea nitida* (rocky soils) and occasionally *Olea europaea* var. *africana* on granitic soils (van Wilgren & McDonald 1992). *Erica* spp. are also dominant in some areas, including *Erica hispidula* and *Erica versicolor*.

Their primary associates include *Restio inconspicuus, Tetraria bromoides* (with *Erica hispidula)* and *Agatho-soma ovata* (with *Erica versicolor)* (McDonald 1993). In riparian areas, 3-6 m high forests composed of *Brabejum stellatifolium* occur, with an understorey of *Diospyros glabra, Halleria ellip-tica* and *Rhus angustifolia*. Tall (8-16m) forests grow on stable boulder screes (van Wilgren & McDonald 1992). Canopy species include *Harto-giella schinoides, Olinia ventosa* and *Podocarpus elongatus*. On loose boulder fields, short forests (2-8 m) consisting of *Hartogiella schinoides, Heeria argentea, Olinea ventosa* and *Maytenus acuminata* grow, with a *Podocarpus elongatus* understorey.

## SUMMARY

The earth supports a tremendous amount and diversity of forests, which humans have utilized as a source of food, fibre and forage for countless millennia. In the process, we have inventoried forests based on their composition and structure, recognizing each for the particular products they supply and the organisms they contain. The inventory and classification process was informal and driven by intuition and necessity in the past, but within the past century or so it has been performed increasingly by design. By virtue of their low species diversity and limited geographical or topographical distribution, some forests (e.g. the coniferous forests of western North America) have lent themselves well to more formal and specialized classification. Others, with abundant species, many of which have highly localized distributions (e.g. the tropical rain forests), have proved difficult or impossible to assign to a particular forest type. Moreover, many economically important species have been widely planted outside their native range, so much so that some (e.g. *Pinus radiata)* are far more abundant outside their native range than within it. In spite of these difficulties, I hope that I have been able to adequately differentiate among, and generally describe, the world's forest biomes and the forest types or communities within them, and at the same time conveyed to the reader an appreciation for their rich ecological diversity.

3

# Genetics and Speciation in the World's Forests

## INTRODUCTION

The vast majority of plant and animal diversity is contained in forests, comprising both the trees that dominate this environment as well as forest-dependent species. The genetic and species diversity of trees provide the foundation for much of the earth's biodiversity. Among the trees themselves, taxonomic diversity is exemplified by the 25000 species and the high levels of genetic variation within most of them. While this is only around 10% of the total number of vascular plant species, the forests of the world cover over one-third of the dry land area.

In addition, the functional diversity of tree species is commonly assumed to be necessary for the sustenance of forests and the organisms dependent on them. An equivalent amount of the dry land area of the earth is in pasture and agriculture, but these ecosystems are relatively low in species and genetic and functional diversity. Hence, much of the world's biodiversity at the genetic, species and ecosystem levels of organization depends on the processes that create and sustain the genetic and species diversity of forest trees.

With the loss of original forests around the world, there is concern over the robustness of this diversity, and a need to understand its origins. How did the trees arrive at their present locations and how did tree species arrive at their present distributions? Do the accidents of history determine the outcome of evolution or is there selectivity for an optimum distribution?

Are some families simply more prone to diversify and to create new species? To what extent does arrival time and the capacity to migrate determine distribution and how much influence is exerted by the capacity to grow and reproduce, and to adapt to local competition for resources?

Forest composition is the result of several genetically influenced factors, including the probabilities and distributions of species establishment and their responses to the stresses and opportunities for adaptation. Genetic capacities thus influence ecological dynamics. However, once established on a site, the levels and distribution of genetic variation are partially determined by stand dynamics and the distribution of available parents for establishing the next generation. Thus, the genetic patterns of variation are partly influenced by establishment, survival and reproductive dynamics, and species dynamics are partly determined by genetic dynamics. Since these factors are particularly complex in forest ecosystems, the relationships between genetic and ecological dynamics are particularly important to understand. The evolution of forests is neither a matter of genetic factors preceding ecological dispositions nor of ecological factors determining optimal genetic composition; it is an interaction of genetics and ecology.

The joint dynamics of genetics and ecology are also influenced by historic events. We might first note that in the evolution of the modern species of forest trees, the history of distinctive species is relatively long. The fossil record indicates that recognizably similar trees have existed for much longer periods of time than for the grasses and that the range of environments over which trees grow is also much larger than for annual plants. This implies that forest trees have had a long period of time to evolve, for some species to go extinct and for others to proliferate variations. It also implies that a large variety of evolutionary histories have had the opportunity to develop over large spans of time and geography.

In a broad view of forest ecosystems, there are about 4 billion ha of forests. Approximately 32% are in the coniferous forests of Europe and North America, 27% in other temperate forests and 41% in dry and moist tropical forests. Since the tropical zones contain the greatest number of species, there is an uneven packing of species around the world in different forest ecosystems. Some families and groups of genera are more heavily represented in some parts of the globe than others. Historically contingent migration and the existence of migration routes also affect where species may have had the opportunity to exist. In addition, environmental differences would give selective preference to some species and hence allow only selected species to compete for resources. A further consequence of differential species packing is that the number of individuals per species must be reduced in any given forest if species number increases, and this would affect the density distribution and mating system of those species. Some species may evolve as environmental specialists that avoid competition and mate in small neighbourhoods.

In addition, historical contingencies as well as selective factors affect the dynamics of within-species distribution. Among species, we would therefore expect differences in how they are subdivided, how distant and how often parents can find mates, and how they adapt to the different environments that they encounter. With different climates and soils around the world, the different species that react to their environments can be expected to develop different evolutionary responses and different life histories for their growth and reproduction. Then, since species dynamics differ, we can expect that the distribution of genetic variation would also be different among the species in different ecosystems. Thus genetic variation is both the product of evolution and the engine that drives the capacity of organisms to adapt. The interplay of environment and history that affects the distribution of species and genes is affected by the genetic and evolutionary capacity of those species and has created the rich biotic dynamics which is our forest legacy.

## THE WEALTH OF FOREST TREE SPECIES

The earliest forests of the Carboniferous Age were probably not composed of woody tree species but of ferns, cycads and other vascular plants that were neither conifer nor angiosperm. The more complex vascular and reproductive systems of coniferous species did not evolve until the Devonian Period some 350 million years ago. Conifer trees were not prominent until the Permian Period 100 million years later (Scagel 1965). It was another 70-100 million years before the pines differentiated, around 180 million years ago. The soft and hard pines diverged around 130 million years ago, during and following the break-up of Pangaea and the great continental migrations. Angiosperms evidently did not originate until the Cretaceous Period, but soon broadleaf tree species would become more prominent in forests. By around 100 million years ago they began to dominate over the ferns and gymnosperms in many forests.

The end of the Cretaceous Period, around 70 million years ago, is marked by major animal extinctions, wide climatic fluctuations, continental shifts with episodes of mountain building, and the differentiation of the major angiosperm family groups and genera. Periods of major extinctions also occurred during the Permian and during the global cooling that took place at the end of the Tertiary. Further speciation continues in both gymnosperms and angiosperms to this day, but largely on the development plans established by the end of the Cretaceous.

Since most of the major families and genera existed at least as prototypes by the end of the Cretaceous Period, there are large patterns of geographical differentiation in species, genera and families. Among the nearly 600 species of conifers, the Pinaceae (10 genera, 250 species) is a Northern Hemisphere family, as is Taxaceae (5 genera, 15 species), Cephalotaxaceae (1 genus, 5 species) and Taxodiaceae (9 genera, 12 species, with one southern genus with 3 species), while Podocarpaceae (7 genera, 131 species) and Araucareaceae (2 genera, 36 species) are predominantly Southern Hemisphere families. Cupressaceae (16 genera, 142 species) is spread over both hemispheres. All species are monoecious and wind pollinated.

Among the angiosperms, a much larger number of species (20000-25000) evolved during the Cretaceous and the development of the progenitors of many presently known species continued through the early Tertiary Period of large climate change and widespread extinction. A wide variety of reproductive behaviour evolved, including monoecy and dioecy, and all forms of sexual divergence in between (Bawa & Opler 1975; Charlesworth & Charlesworth 1978; Charlesworth 1984, 1989), which affected the rates of population and species divergence and evolution.

However, the establishment of mating barriers is not the *sine qua non* of speciation for plants as it is for animals, and other forces have relatively more effect on the creation of new species (Grant 1971; Grant & Mitton 1979). To understand how the present array of tree species arose and how future diversity may evolve, several evolutionary factors must be considered.

The selective forces of the environment may favour different inherited traits and force divergence. Mutations may also generate diversity, and the reproductive system itself may erect barriers to intermating among the derived and the progenitor groups. Other events such as geographical isolation may form barriers, and further isolation could reinforce the effects of selection and mutation to generate distinct species. If further divergence occurs, then species-level differences can accumulate to generic and familial levels of divergence as later speciation events accumulate differences and sets of descendants diverge further. Such divergence may be rapid or slow depending on the strengths of the evolutionary factors, but all factors were probably strong simultaneously during those periods of rapid speciation such as occurred at the Cretaceous-Tertiary boundary.

## Cladogenesis

The development of species (cladogenesis) is seen as a process that proceeds from the normal evolution of a species which, if no split occurs, would be called anagenesis, or the evolution of a species lineage without subdivision. Within a human lifetime and on a local scale of observation, microevolutionary changes composed of shifts in species traits can be observed, such as when tree breeding programmes change growth rates or stem form. Species seem to remain fairly constant in their general form and behaviour, though with gradual shifts in a continuing evolution of adaptation (anagenesis). Evolution may be largely dominated by events common to the whole species, if populations are often intermating and subject to similar environmental conditions. Even if subject to different and strongly selective environmental conditions, occasional interpopu-lation mating can force their evolutionary dynamics to be very similar to that of a single population. However, on a longer time-scale than included in a single human generation, or on a large geographical scale, species may not always evolve as a unit with gradual continuous steps.

On a global scale, it is expected that environmental differences are so large that the segregation of different adaptive capacities would often be advantageous and hence that selection on heritable traits would be a factor in speciation. Since ecologically critical factors are not distributed uniformly around the world, divergence rates among species would also not be expected to be uniform. We can see some of the broad strokes of large effects in the contrasts between the boreal and tropical forests. The obvious factors of temperature and moisture are so different among broad regions that divergent adaptations, at least to major climate extremes, would clearly be advantageous. In addition, it can also be observed that in the boreal forests most species carry both sexes on the same tree, have their pollen carried by wind and are usually found in relatively large continuous stands.

In contrast, tropical forest trees often carry males and females on different trees. These species are most often sparsely distributed throughout most of their range (Ashton 1976; O'Malley & Bawa 1987), and insects, birds or mammals carry pollen. Opportunities for inbreeding or for restricted localization of mating neighbourhoods are much greater in such species (Bawa 1974). Tree species differ in their reproductive behaviour as well as in the environments to which they are better adapted. It can therefore be expected that forest regeneration dynamics would differ among regions and that their genetic dynamics would also differ, leading to segregation into divergent populations or subspecies.

Among the angiosperms, many reproductive systems exist, from those that are primarily self-pollinating to those which are primarily outcross-ing (Murawski & Hamrick 1991). Pollination may be predominantly mediated by wind, insects, birds or mammals, and seed dispersal may be proximal or distant; both may be episodic between years or may be continuous throughout a year. The mating system of some species is conducive to restricting the pool of potential parents. The mating system may also force some species to depend on one or other type of pollinator and would lead to reproductive isolation (Bawa 1974). In addition, the possibility of the mating pool being very limited would, if coupled with a non-continuous ecological distribution, allow some species sets to have very rapid rates of evolutionary divergence.

Thus if ecological conditions act as selective agents that differentially affect survival or reproduction of some genotypes, the genetic composition of the population may be biased by selection for tolerance to competition or by sexual selection on rare genotypes. In such cases population evolution would be dependent on the combination of genetic and ecological events. The evolution of populations can then affect not only the immediate survival of that population but can also determine the future subspecies structures that may lead to further speciation or other genetic structure for the species as a whole. Thus, the mating system, environmental differences and history can all affect the rates of species creation by cladogenesis. Cladogenesis and anagenesis would be expected to often follow the same rules of evolutionary dynamics whether in single or multiple populations, and to be determined by the effects of mutation, migration, selection and population size.

The forces of evolution act on the capacity of a species to respond to challenges of the reproductive or vegetative environment, and that capacity is determined by its genetic endowment. This genetic endowment determines what it is that we call a species at any one moment in its evolution and also determines what its evolutionary trajectory may be. Therefore, discussing the species present in forests requires discussion of what their genetic endowment is, as well as what forces are acting on it. Discussion on how species are changing, or can be conserved, also requires discussion on how their genetics is changing and how their genes can be conserved. Since many of the dynamical events that affect evolution occur among individual trees, the individual is considered to be the primary unit of evolution, and the genotype (the collection of all the genes in an individual) is the unit of gene expression. The phenotype is the physical expression of the genotype within the environment of the individual's lifetime and is the unit of selection. The correspondence between individual-level events that we can observe and the long-term evolution of species groups, which takes a longer time, is the realm of evolutionary and population genetics.

However, evolution is not a unidirectional phenomenon, with trees and their genes passively exposed to a selective environment. It does not proceed solely from a given set of environments that determine which genes are present, nor solely from genes to how the trees will perform. At a large level, it is clear that the forces which affect reproductive behaviour and selection are themselves changing partly as a result of the genes that are favoured. As climates and human influences change, the ranges and density distributions of species can also be expected to change. The migration of species may thus be driven by general environmental influences, but would also be influenced by how other elements of the biota abet or hinder their movements. These are factors that are partially dependent on which genes are present. For example, in the temperate and boreal forests the postglacial movement of species would not be expected to be uniform since they differ in colonizing ability and their initial distributions (Pielou 1991).

In addition, some species would form migrational barriers to other species and hence the history of migration would also affect the distribution of species and would inhibit the existence of an optimal composition of species in communities. One pioneer species will often exclude other potential pioneer species, and secondary succession may depend on those initial events as well. Thus, tree communities may in fact be associations of species that have not coevolved to their mutual benefit but coexist due to independent adaptations to similar environmental conditions, or that jointly occur for ephemeral and historically dependent reasons. Evolution within forest tree species may therefore be largely independent of the evolution of other tree species. They are less likely to be independent of interacting species such as their pollinators.

The dynamics of forest ecology depends on the characteristics of the species that can invade and maintain themselves, and is therefore partly a result of the species endowment available. However, the evolution within species also depends on those same ecological factors. While it may be obvious that major differences in forest dynamics exist among the forest zones, all forests go through periods when some species establish themselves in new populations while others die out, at least in local areas.

On a given area, the displacement of one species by another is a common occurrence that can be triggered by extrinsic forces such as fire or by autonomous factors such as competitive exclusion of one species by another (Shugart 1984). Some of these displacements are mediated by differences in species adaptations to regeneration under shade vs. open areas or to differences in competitive ability for soil, water or light resources. Some displacements may be due only to the accidents of the timing of seed arrival or other random events.

While a forest cover may exist in some areas for many centuries, at least local variations in species composition over time would be the general rule and hence the local regeneration of any one species iswould be expected to be often sporadic and episodic (Loveless & Hamrick 1984; Shugart 1984). Even for forests composed of shade-tolerant and self-replacing species, gaps frequently disrupt the continuity of forest cover and generate sequences of displacement (Clarke & Clarke 1984; Clarke & Clarke 1991). Some species, called pioneers, are also dependent on large gaps or cleared openings for regeneration, and normally form dense colonies in the foundation of populations (Bannister 1965). The seeds of such species may be easily transported over long distances (Wright 1976) or may be stored in a soil seed bank, enabling cleared areas to regenerate a forest cover. Their mode of regeneration also implies that they suffer local extinctions, through their own effects on their environment, through displacement by superior competitors or by environmental catastrophes. Forest dynamics itself brings evolutionary pressure on the forest's constituent species.

## Selection

The selective advantage of one individual over another is one of the main forces of evolutionary genetics, and along with mutation, migration and drift form the genetic endowment of a species and of the populations of which it is composed. Evolution from the individual level of selective and reproductive events is one of the ways that we can understand the transition from what is called microevolution to higher taxonomic levels, or macroevolution.

Competition dynamics is often invoked as one of the strong selective factors in forest tree evolution (Campbell 1979). Species that benefit from or tolerate shade as seedlings or saplings may occur in higher frequency under a canopy, or may invade areas already occupied by other vegetation. They may emerge into the upper canopy when openings occur and hence the distribution of reproduc-tively mature individuals would follow patterns established by the size and frequency of canopy openings (Bawa 1974). Some species in tropical forests require shade or a protective canopy cover and can reproduce only under a canopy where their crowns can form a low canopy layer. However, light is not the only environmental limiting factor that affects the distribution of trees. Other factors such as soil type and soil moisture can have a strong influence on the extent and patchiness of a species distribution.

One of the basic necessities for evolution is the capacity to adapt to new stresses in the physical or biotic environment (Loehle & Namkoong 1987). For any selection to lead to adaptive evolution, there must be genetic variation in the stress response and the trees with the favoured trait expression must leave more progeny.

Since many environmental factors are themselves changing in response to extrinsic forces, we can expect that stand composition and species distributions are perpetually changing for extrinsic reasons. Thus, selection factors are not expected to be consistent over time. In addition, historic influences on the recruitment pool for species establishment can affect the probabilities of species replacements and simple stochastic variations on the demographic distribution can also influence distribution for reasons not related to optimality of fitness. Hence, the accidents of history, stochastic variations in environmental variables, lag times in selective effects, as well as deterministic causes affect the actual distribution of species.

Virtually all species can be considered to be involved in a dynamical system, to still be migrating both geographically and ecologically, and to be evolving in response to directional evolutionary forces as well as to both deterministic and stochastic forces. Therefore, forest tree species can be expected never to be in an optimally adapted state or, if so, only transiently.

Thus, while selection can be a significant force in the evolution of species and in the evolution of diversity among species, it has effects within a complex ecosystem and on complex organisms, and its effects have a lag time. Its effects also depend on the genetic endowment and structure of variation already present in the species and on the other forces of evolution, including mutation, migration and drift. These forces can change the species characteristics and if sufficient genetic variation is maintained within or between populations, continued evolution can maintain the species and can sustain forests. *Sustainable forestry and the maintenance of species thus require a sustainable genetic resource in some form.* The wealth of genetic variation within and between populations and the combined genetic and ecological dynamics thus jointly determine the inherent structure of forests.

## Migration

If we examine how species migrate and change their geographical distributions over a few generations, the establishment and extinction of local populations is not an unusual event. If such populations become reproductively isolated and if, either by chance or through selective events, they also genetically diverge in mating behaviour and adaptability, the basis for developing a new species exists. If populations do diverge, they would be likely to do so at many genetic loci and at many alleles at the same time. Hence, whole suits of traits and sets of genes would be expected to shift together as populations segregated into different mating groups. Not all of these changes would be ascribable to selective advantage of the alleles, but rather to chance events and to accidental associations of genes and traits.

In the relatively simple boreal ecosystems, the movements of populations escaping from glacial encroachment (Pielou 1991) into sufficiently hospitable areas and alternately advancing into retreating glacial boundaries would be expected to allow populations to carry adaptations to conditions that lag behind actual environments. Nevertheless, during episodes of environmental change or of species movement into new environments, some traits would be of adaptive significance and near optimality, while others may be of little significance or may lag far from any current optimality (Namkoong 1969). Only if the populations carry sufficient genetic variation could they respond rapidly to optimizing selection and ultimately achieve or approach an optimum condition, and then only if the environment is also stabilized.

During periods of change, populations may diverge from a base population or from each other in location or habitat (Grant 1971). For divergence to become inherent among them, strong selection on populations with genetic variation in traits of adaptive significance and an almost absolute barrier to pollen or seed migration between populations would be required. For *Pinus contorta,* strong interpopulation differences can be observed despite substantial gene migration (Yang & Yeh 1995). Most often, it can be expected that without designed or accidental interventions, separated populations would sooner or later reconverge and that periodic intermating among populations would be frequent enough that traces of different gene frequencies and linkages would be difficult to detect. If populations once diverge, they may often simply integrate upon secondary contact and re-form as a single intermating population. Without strong mating barriers for a long time, the long-term evolution of the species as a whole would be a slow approach to an average optimality by the increments of mutations to new alleles at some loci.

Occasionally however, divergence is strongly established for long enough that at least partial mating barriers are formed. Then, splitting of a single former species into two or more branches could generate new species. While the occasional speciation event may not be often repeated, incipient speciation could generate stronger and stronger divergences in genetic composition as selective differences and unique mutations accumulated. Once a coherent species is formed, any further speciation events would follow on the basis of prior divergence and there could then ensue a hierarchical series of branching species that ultimately develop into genera and family level divergence. This is the basis on which phylo-genetic relationships are constructed and which underlies the power of molecular genetic data for reconstructing evolutionary trees.

The existence of large populations undergoing population isolation and of changing environments were pronounced during the Cretaceous Period, when many of the major forest tree speciation events occurred. Subsequently, this led to the generic and family level structures that we see today. Though the potential for speciation events continues to exist through modern times, the events occur at slower rates than humans are attuned to observe. However, by using experimental evolutionary approaches, including strong selection and the construction of strong mating barriers, such as among divergently selected breeding populations, it is possible to create large population divergence both genotypically and phenotypically. As long as new species arise by splitting parental species, and each accumulates random mutational changes, divergence among species would follow patterns of statistical distributions, and by tracing backwards in time, similarities can be used to reconstruct coalescent histories.

## Chromosomal change

Occasionally speciation events in plants also occur as a result of reorganization of the genetic material, especially by means of chromosome rearrangements and duplications. For example, it is reasonable to conjecture that chromosome rearrangements explain the number of chromosomes of Douglas fir, which has 26 chromosomes in its vegetative cells instead of the usual 24 carried in all other conifers. This is probably the result of one of the usual conifer chromosomes dividing in two and developing its own centromere. In tropical hardwood species, it is not uncommon to find irregular series of chromosome numbers (Mehra & Bawa 1969), seen in the dipterocarp species especially (Jong 1976).

Within most families, the basic haploid number of chromosomes ($x$) is the same for all genera and species, but for some, as among the Aceraceae and Betulaceae, there is a wide variation in chromosome number among genera and species (Grant 1971; Morgenstern 1996). There is a series of species with chromosome numbers that are multiples of the basic number (e.g. $2x$ to as much as $6x$) which form distinctive species. In aspens, occasional triploids arise following species hybridizations (Morgenstern 1996). The genes in these species may simply be replicated among the chromosome sets and all continue to add to gene expression.

However, the kinds of interactions among alleles can be affected by their pairings as well as by the numbers of alleles at the same 'locus' and among 'loci'. Such a series of chromosome multiplication is a polyploid series, and may arise as the result of a meiotic failure of the original nuclear genome to split into two separate nuclei. This may occur often enough within a population that the pollen or egg cell can mate with another that also carries the doubled set, and instead of producing a diploid would produce a tetraploid. If a diploid germ cell is paired with a normal haploid cell, a triploid would form, although these usually fail to generate viable germ cells. However, if a diploid and a tetraploid happen to mate and those cells double their chromosome set, a hexaploid could result.

The creation of polyploid series may be stimulated by hybridization events, where chromosome mechanics are sufficiently disturbed that unusual meiosis events may occur more frequently. However, hybridization events by themselves can also create new kinds of species dynamics such that new species are generated even if the chromosome numbers are the same and do not otherwise change.

## Hybridization

Hybridizations among divergent species occur when they are not too divergent for the intermat-ing to occur but are distinctly different in form or function. They may be frequent enough among plant species that their potential for shaping evolutionary relationships cannot be ignored (Barton & Bengtsson 1986). Hybridization occurs with much greater frequency among plant species than among animals. Examples of hybridization in fruit trees can be found with different outcomes. At some point during the divergent development of distinct species, a barrier may exist that further reinforces intermating such that no hybridizations may occur, although this seems to be at fairly wide levels of divergence. Before that time, hybridizations can intermingle the genetic endowments of the incipient species and can thereafter serve as foundation taxa for further evolution of divergence among species. Therefore, the existence of hybrids and populations of intermediate genotypes implies that the speciation process has not gone so far that new species can evolve by hybridization. In fact the processes of divergence and hybridization are continuing to this day. The large areas over which lodge-pole pine and jack pine hybridize indicate a broad compatibility of distinct species that may stabilize in some areas (Wheeler & Gurries 1987), but with intermediate types developing as a new taxon.

In some cases, the population swarm that characterizes recent hybridizations may segregate the parental types as second-generation intercrosses or as backcrosses to their parental populations.

Alternatively, they may maintain intermediate types in most traits and eventually stabilize as a taxonomic unit with some intermediate trait expression and with some mixture of different parental traits.

In the evolution of a hybrid population between *Pinus taeda* and *Pinus palustris* created by the natural intercrossing of the parental species in the 1920s, later generations failed to maintain any intermediate types and, 20 and 40 years later, when their progeny reproduced they predominantly gave rise to the parental types (Namkoong 1966). Apparently, the original hybrids were successful only in the peculiar environment of abandoned farms from which fire was excluded; when a forest cover was established, with occasional brush fires, the hybrid either was not selectively stable or was chromosomally unstable, or both. In contrast, the hybrids between *Picea sitchensis* and *Picea glauca* form a large and apparently stable zone of introgression in northern British Columbia and hence can be stabilized as a hybrid with backcrossing to both parental species (Yeh & Arnott 1986). The hybrid types may disappear or may form a permanent zone for hybridization, or may eventually form a third 'species'.

If just two parental species are involved in the development of a hybrid, it might be difficult to distinguish if an intermediate type is a recent hybrid of historic species or if the intermediate is a historic source population for newly divergent taxonomic units. If historical evidence is available on prior migrations, then the origins of the units can be readily distinguished by the times of establishment. However, in the absence of historical data the intermediate may be either a hybrid progeny of two divergent parents or the parental source of divergent progeny. If genetic measures can be obtained for multiple, independent and neutral loci, it would be expected that the isolation of divergent populations would initially involve a random segregation of genes from the original population.

This is a case of single source population producing two derived branches, and instead of simply splitting into two, three branches may exist from a single origin. Then the derived progeny populations would receive an admixture of alleles such that among the progeny populations themselves there would be similarities in some traits and differences in others. If alleles or traits were different between the progeny populations, they would have diverged from the parent population but would not necessarily diverge in symmetrically opposing directions. The allelic richness of the source population would be expected to be at least as great as that of both of the derived progeny populations, each of which would contain subsets of the parent population's alleles.

In contrast, a recently derived hybrid population would share most alleles at frequencies between the parental populations and would therefore be intermediate in most traits simultaneously. The divergent populations then would seem to diverge in symmetrically opposing directions from the intermediate population at all divergent loci and in all traits. The hybrid population would also be expected to have alleles that exist in at least one of the parent populations but not to contain all of the alleles of both parents.

The eventual divergence of hybrid populations can, of course, involve backcrossing and selection such that they ultimately cannot be distinguished from the previous model of divergence from a parental stock. However, linkage disequilibrium in the hybrid population should linger and be detectable for many generations. In this case, two parental species can give rise to three species if the hybrid continues to develop and establishes mating barriers such that it can evolve independently. In the case of *Pinus rigida, Pinus taeda* and *Pinus clausa,* there are three two-way hybrids between the species that could indicate a close relationship between them and a splitting of an original one into two or three by a subsequent split. However, it could also imply a recent hybridization of at least two of the parental species (Smouse 1972).

These forms of speciation can lead to further diversification of lineages or to convergence of genetic lineages, if some sets of genes are transferred among several species. These forms of speciation may then form multibrachial taxonomic structures and cross-linkages of sets of genes, neither of which conforms to evolutionary trees. The existence of three-way hybrids in pines, for example, indicates that such complexities in speciation may not be unusual at least among the pines and may not have been unusual in the development of the family. The construction of genetic lineages therefore is not identical with the construction of species lineages, and local events in the evolution of populations can have significant effects on long-term phylogenies.

The events involved in speciation do not generally occur in all units of a species at the same time, but instead involve genetic events that change individuals or that occur among population isolates. If species-level events are contained within a matrix composed of other species, they may often exist as a set of reproductively isolated islands, and if the parents of new populations rarely mate with members of other islands, the species may exist in a metapopulation structure of loosely connected subpopulations. This may even lead to incipient speciation as the islands become more distinct if they further form mating barriers among populations. Thus, a species may be distributed in the form of large, indistinct, continuously intermating populations, or as discrete and distinct metapopulations and in mixtures of large and small populations with partial migration among some subsets. The amount and distribution of genetic variation will be influenced by the metapopulation structure as populations split into large or small segments and as they either advance or retreat from habitable sites.

## THE WEALTH OF POPULATIONS

Within plant species, genetic variation commonly exists at a high proportion of loci and tree species commonly exhibit the highest levels of all plants (Hamrick & Murawski 1991). For the species surveyed, which are mostly boreal and temperate pioneer species, the levels of genetic variation within populations often seem to exceed those between populations. The lack of strong differentiation may be due to the wide dispersal abilities of pollen and seed, which maintain a degree of connectedness between mating neighbourhoods among trees over long distances. A few studies of tropical tree species that are not pioneers indicate considerable variation among populations of trees, which may reflect a finer subdivision of mating neighbourhoods.

Another measure of genetic variation is the heritable variation that exists in traits controlled by many gene effects simultaneously but in which individual gene loci may have small effects. These are the so-called quantitative traits that can involve many genetic loci which act indirectly on the observed trait and hence have small direct effect. Most of these loci would not change much in frequency from one generation to the next. Cumulatively, however, they could cause large changes in the trait expression. Thus, many of the traits selected for in breeding populations can shift mean expression by as much as one standard deviation in one generation even if there is little shift in gene frequencies at loci sampled by molecular means. Furthermore, since differences among populations in the frequency of molecular markers are more difficult to detect than differences measured in quantitative traits (Lewontin 1984), traits of adaptive significance may be more obviously segregated among populations than molecular genetic measures may indicate.

Some traits and the gene loci affecting them may be under strong selection pressure that is divergent between populations, while other traits may be under mild or virtually no selection pressure. Some traits may be indirectly related to strong selection events, such as caused by fire, drought, or forest practices that leave few resistant individuals, but only weakly related to resistance mechanisms. Other genes may not be presently involved in any recent selection events. Thus the genes associated with selection response may be under strong or weak selection according to the strength of the selection event and the degree of relatedness of the trait to the selection agent. There may also be multiple genes that affect a trait expression, and the combined effects of the genes may not be simply assigned to any simple action of any one of them. This kind of multigene interaction is called epistasis and it may cause multiple gene locus differences to be more pronounced at the level of trait observations than in single-locus allele frequency differences. These kinds of interactions may mask any obvious trends in gene frequency changes associated with selection trends.

Since the phenotypic traits themselves can be expected to be under a wide array of selection intensities and the genes associated with them can also have a wide range of influence on those traits, we can expect that there would be a wide range of selection pressures among the genes of any one population (Campbell 1987). Thus, if the relative divergence of populations was tracked by weakly selected traits or loci, population similarities would predominantly reflect migration effects. In the absence of selection, even very low migration rates would tend to maintain similar allele frequencies among populations. However, if divergence was tracked using traits or loci that respond strongly to divergent habitat selection, then similarities would tend to reflect habitat differences. In any one population, strong directional selection would force disfavoured alleles towards the point of extinction unless maintained by an input from migration or mutation.

Among populations exposed to divergent directional selection, different alleles would be at high and low frequencies among them and hence their frequencies would be highly dissimilar. If a large number of traits or loci are used and some loci reflect similarities where others reflect multiple patterns of divergence, then contrasting patterns of allelic distribution must coexist. Similarly, traits subject to different intensities of divergent selection would also display different degrees of divergence. Clearly, if phylogenetic relationships were estimated using selectively significant loci along an environmental gradient, then genetic similarities or distances would be correlated with that environmental variable. Alternatively, if selectively neutral traits were used, the genetic distance measures would be correlated with migration distances, and the derived phylo-genies would be different. The existence of differing strengths of divergence has been observed in variations in the patterns of geographically related population traits in *Pinus armandii* (Ma 1989). Differences in the amount and geographical pattern of isozyme allele frequencies and in growth traits have also been found for *Pinus lam-bertiana* (Martinson 1996) and red alder (Hamann *et al.* 1998).

In addition to those forces of evolution, the mating system can have heterogeneous effects on the distribution of alleles. If pollen and seed have different migration distances or different direction of flow due to their migration season or the influence of different wind pollinator effect on the sexes, then the balancing effects of migration and selection would be different (Namkoong & Gregorius 1985). Even for selectively neutral alleles, differences between the migration paths of the sexes causes heterogeneity of allele distribution.

Genetic variation also exists at several levels of expression and the amounts and patterns of variation can depend on which measures of variation are used to estimate that differentiation. With currently available tools of molecular measurement, it is possible to measure very fine levels of difference so that individual trees (or humans for that matter) can be discriminated by unique changes in the fine structure of single genes. It is also possible to discriminate differences among local populations if they share maternal or paternal lineages, by using particular molecular genetic markers that distinguish among populations even though the populations are not obviously different phenotypically (Millar & Libby 1991). Furthermore, some genes are expected to be more influenced by the different forces of evolution and therefore the distribution of variation among genes may vary according to their function (Ma 1989).

Therefore, there is no one preferred measure of genetic variation but several options, and the choice of which measure to use depends on the purpose of the measure and the type of gene being measured. For forest trees it is satisfying to note that the levels of genetic variation measured are usually among the highest in the plant kingdom, although their distribution and the differences among loci have not been well studied. If the frequency of heterogeneous associations is high, the evolution of forest species may be unusually complex.

Some forest trees have relatively high levels of genetic variation between populations in several traits and at several molecular loci, while others may have low general levels (Millar & Libby

1991). Since the beginnings of interest in genetics, foresters have known about economically significant levels of heritable genetic variation among populations in traits relating to growth and soil and climatic adaptability (Morgenstern 1996). One of the earliest tests to indicate the utility of population differences is recorded for *Cryptome-ria* in Japan in 1883 (Toda 1961). However, the existence of graded changes in traits is widely recognized in forest trees as clines (Teich & Holst 1974; Millar & Libby 1991; Morgenstern 1996) and with high levels of variation among populations relative to genetic variation within populations (Wheeler & Gurries 1982). The variation patterns often differ depending on which traits are measured but growth rates and climatic adaptabilities are often considered to be optimized within local populations (Rehfeldt 1989). While there is debate as to the generality of the concept that local populations are optimal (Namkoong 1969; Rehfeldt 1995), regular trends in performances are often observed.

The geographical patterns of variation among populations or provenances may vary from a single, large, contiguous and intermating population with gradations across mating and environmental neighbourhoods to one of discretely divergent populations with little relationship among subpopulations. Some species may be composed of a main, large, central population with satellites connected only to the central one and with only one-way germplasm flow, or may be a series of more or less equivalent-sized populations with equal and multidirectional flows. Some of them may exist along streamside corridors as single units with connections between drainage basins being more tenuous, while others may form island populations on the tops of separate mountain peaks or ranges.

Since the stability of the interpopulation structure depends on the stability of ecological dynamics, the level of population or provenance differences may fluctuate over time and geographical area. In turn, the response of populations to environmental divergence is contingent on the genetic dynamics that can occur within an intermating neighbourhood. For the most part, the potential for selective divergence among the populations is moderated by migration among populations. Only traits with strong divergent selection would maintain high levels of interpopulation difference.

## THE WEALTH OF INTRAPOPULATION GENETIC VARIATION

Within populations, four forces (selection, migration, mutation and population size) determine the structure of genetic variation. The interaction of all four factors can be highly complex and their effects are determined by the mating system and the kinds of gene effects that exist. A problem in analysing even this level of genetic evolution is that the kinds of gene effects that exist depend on the effects that other forces of evolution have had on gene actions themselves. To consider how the forces can affect evolution, we can first start with each of them and then consider how their interactions can change predictions.

For selection to have an effect on speciation and the structure of forests, there must be a difference among genotypes in the traits affected by survival or reproduction events and the individuals must accordingly leave more or fewer progeny than other genotypes. The genotype is defined as the individuals with a distinctive ensemble of alleles at its gene loci that distinguishes it from other genotypes. This is a multiple gene locus definition, which is often simplified by considering every one of the loci to be independent in the occurrence of its alleles in both frequency and effect.

In this case, we can consider the genotype to be the summation of all independent loci and hence can examine gene effects and dynamics one locus at a time. This is, of course, an oversimplified view since genes are in fact linked on chromosomes and therefore have linkages that correlate their occurrences, and most genes affect processes that are combined in their net effect on traits; furthermore, most genes affect enzymatic pathways that ultimately affect several traits when measured at the level of whole organism performance. These effects are called linkage, epistasis and pleiotropy, respectively, and tie the whole organism together into an integrated set of actions and functions that can survive and reproduce. While undoubtedly true in general, there is little evidence that individuals are so bound by such constrained genetic constitutions and physiological limitations that substantial independence among loci exists. The evidence for the existence of strongly selected, tightly linked 'coadapted gene complexes' is not strong and hence the model of independence can be usefully employed, even though it is surely wrong at least in detail.

Within a gene locus, many alleles may exist and the pairwise interactions among them are called the dominance relations. For a diploid organism, if both alleles at a locus are the same, the individual is said to be a homozygote; if the two alleles are different, the individual is said to be a heterozygote. If the heterozygote has a phenotype intermediate between the two homozygotes, the dominance effect is considered to be zero and the gene action is said to be additive. If the heterozygote is other than intermediate, it may be less than the lesser of the two homozygotes. If its phenotype lies between the homozygotes but is not exactly intermediate and if it appears to be greater than the greater of the homozygotes, the gene action is said to be underdominant, partially dominant or overdominant, respectively.

The effect of selection would depend on the kind of gene action, since directional selection for greater or lesser phenotypes would favour different homozygote and heterozygote states. Selection may also be stabilizing if it favours an intermediate phenotype or disruptive if it disfavours the intermediate and its effect on the frequency of the alleles would depend on the gene actions (Namkoong & Koshy 1995). Directional selection experiments seem to indicate that most alleles have partial dominance actions, though many may be involved in multiple locus interactions. Since selection would be expected to favour survival and reproduction of certain genotypes, we expect that it would increase the population level of fitness.

Selection is therefore usually considered to have a positive effect on the evolutionary potential of a population or species in small increments every generation but may not generate an optimal condition at any one time. For alleles that have partial dominance effects, the dynamics of the rates of displacement of one allele by another are such that the frequency of the allele itself affects its own rate of displacement. When alleles are at either low or high frequency, their rates of change are slower than when they are at an intermediate frequency. Except in small populations where accidental loss can occur, selection would not generally lose alleles that are favoured for adaptation.

Another force of evolution is mutation, which is usually considered to be a random change in gene structure and function that occurs indiscriminately throughout the genome. Since most species have evolved a genetic and physiological level that is close to a sufficient capacity for continued evolution, novel changes are not expected to improve performance or fitness but rather to generally decrease adaptability. However, there are always going to be mutations and while most may be disfavoured by selection, there is likely to be a balance between the rate at which new mutants are introduced and the rate of their elimination.

Most mutations occur at low rates for any one locus in any one generation and would not be expected to affect the adaptability of a population unless they are accumulated in small populations. For most conifer species where the frequency of deleterious mutants has been estimated, there are a larger number of lethal alleles carried at one or another locus in most individual trees than in most other plant or animal species. This might indicate that there are many effective loci in trees that can mutate to a harmful state, that the mutation rate is high or that the mutants are favoured in some conditions or are of some advantage in heterozygous conditions (Bishir & Namkoong 1987). However, independent mutations and selection can generate differences among populations, which can increase fitness in each. Hybridization among populations can then generate high levels of genetic variance in traits that may then be useful in new environments (Koshy *et al.* 1998).

A third force of evolution is migration, which can occur among many different kinds of population structures. The input of pollen or seed from other populations that have different allele frequencies may bring in alleles that might otherwise be lost, and if the migration rate is high enough among all populations may induce all populations to have approximately the same frequencies. If the immigrant alleles are favoured, then further immigration would advance the rate of its increase; if disfavoured, immigration would maintain alleles at higher frequencies than expected in a manner similar to mutation; and if neither, would tend to make frequencies homogeneous among all populations. In the presence of divergent selection among populations that exist in different environments, selection would act to eliminate migrant alleles and would maintain population divergence.

The fourth force is sampling error induced by small population sizes, which tends to increase the probability that low-frequency alleles are lost by chance. If populations remain at low sizes for several generations, the chances of random loss are accumulated; even populations of more than 20-50 adults can lose alleles that may have started with moderate frequencies. Since most forest trees seem to carry many deleterious mutants, even if at low frequency at any one locus, another problem is that small populations will suffer from inbreeding depression and may eventually suffer such debilitation from several mildly depressive mutants that the population goes extinct (Lynch *et al.* 1995). Individual populations would have to be large enough to sustain viability, and to avoid environmental accidents and catastrophes if no migrants were recruited.

There are two problems faced in persistently small populations, the loss of alleles and inbreeding depression; hence, larger population sizes are generally better for avoiding extinction. However, the effects of selection, migration and mutation are confounded with the sampling errors incurred in real populations of finite size. Thus, if populations are separated and selected for different environmental optima, migration can introduce alleles in high frequency that depress fitness. On the other hand, migration can maintain alleles in populations where they might be useful, but would be lost due to sampling accidents in small populations.

It is reasonable to expect that all species are under the influence of several or all of the forces of evolution simultaneously and that the balance between the several forces is not the same for all genes in all populations. Furthermore, since the physical and biotic environments of forests are rarely at an equilibrium, the genetic and ecological dynamics of most species would have to be considered to be in other than a stable state.

The actual distribution of the multiple functional forms of alleles at each of the several tens of thousands of gene loci in each of the individuals of a population must therefore be the resultant of the mixture of forces felt at each locus, subject to historical events. The vast possibilities of mixed distributions are overwhelming, although it is also obvious that for most of the species we can study, species are not random mixtures of an infinite number of possible combinations.

## CONCLUSIONS

The wealth of genetic variation that presently exists within species can also be used to diversify populations when subject to intensive selection and breeding. The examples of one- and two-generation breeding effects are numerous and indicate that ordinary breeding in trees can mimic the rate of progress experienced in agricultural crop species.

The early experience with many species indicates that differences in average performance can produce plants that lie outside the range of present performance within three generations, and that within 10 generations there would be no individuals in either population that would overlap in performance with the other type. Within 20 generations, long-term selection can generate even further differences due to the effects of unique mutations in different populations and thus entirely unique gene combinations can exist when only one variety formerly existed.

The beginnings of speciation may therefore not require very long periods of observation if strong selection in populations is coupled with a complete barrier to interpopulation migration. However, if this is possible, it is not clear that such events occurred, or could have occurred, often in the past or that future speciation can be artificially induced.

It is also not clear whether the level of speciation that we observe today has or has not completely explored the space of all possible gene combinations. It may well be that the complexities of growth and development are such that newly desirable trait combinations are not possible within the constraints of development. Similarly, the complexities of linkage and epistasis may impose constraints that prohibit reaching those evolutionary states from any present state. We do not know what our limits are, either retrospectively or prospectively.

The nature of genetic variation within the presently existing tree species is not a fixed endowment, neither is the array of forest tree species that we see today. It is certainly wise to avoid losses when possible, although in a dynamical system losses are inevitable, whether caused by human interventions or not. The evolutionary processes that generate the genetic variation within and among species are in a dynamic state and the same processes are what we use for breeding and which serve as the basis for response to future evolutionary pressures. The question for conservation and management is therefore not what we must preserve but rather what our endowment is and where it will go.

# 4

# Structural Dynamics of Forest Stands and Natural Processes

Forest structure is constantly changing in accordance with its internal pattern of growth and mortality and in response to external forces. Even in natural forests, stands may be 'felled' by natural events, such as storms and fires, whilst stands change between events at a measured pace as individual trees grow, compete, dominate and die. The events that disturb the apparent stability and orderly growth of the forest occur frequently enough to form a permanent influence on forest structure and composition.

This chapter reviews these disturbances and describes their influence on forests at various spatial and temporal scales using mainly European and North American examples. Inevitably it overlaps with other chapters. The 'ordinary' growth of undisturbed forest is considered in Chapters 8 and 9, while Chapter 6 discusses how pests disturb forests. The disturbance regime of any particular forest is to some extent a reflection of climate (see Chapter 11). Whilst most disturbances are localized, some act on a grand scale, influencing forest patterns over an entire landscape. Oliver and Larsen (1996) give a comprehensive account of forest growth and change. Peterken (1996) provides an illustrated summary for the temperate region that is related to conservation issues.

## WHAT COUNTS AS DISTURBANCE?

When a forest that has stood, apparently unchanging, for as long as people can remember is blown down overnight there is little doubt that the forest has been 'disturbed'. The change is abrupt, the event is discrete and the effect on the forest is a complete discontinuity with the previous state of affairs. Although there is always a degree of air movement, it is possible to define, quantify and delimit storms in space and time, and thus to count and study their frequency, pattern and impacts.

When the deer in a forest become so numerous that they browse down all seedlings and basal shoots, the state of the forest is clearly being changed but it is more difficult to see this as a disturbance. The changes are gradual, and in any case herbivores are a natural component of the forest. Abrupt increases or decreases in populations may have immediately discernible effects that can be dated, but more often populations fluctuate over years. In this case the impact of the deer is chronic, a continuously operating site factor, rather than a discrete event.

When a large canopy tree in an old-growth forest becomes rotten at the core and hollowed out by fungi, it is likely to collapse in a breeze that would have offered no threat when it was sound. The chance that a breeze will disintegrate a large tree increases when the branches are covered in ice, the crown is in full leaf or the soil has been soaked by weeks of rain. If one looks at the forest as a whole there will always be a few trees collapsing in this way each year, part of the perpetual cycle of regeneration, growth, death and decay. From this perspective, such mortality can hardly be regarded as a disturbance without destroying the meaning of the word. However, from the perspective of the patch of ground under that tree, the sudden disintegration of a tree that has stood for perhaps 300 years is indeed an event; at this smaller scale it can readily be regarded as a disturbance, which presents an opportunity for regeneration.

The cause of a disturbance is less certain the closer one looks. In the previous example, the cause can be the great age of the tree, the presence of fungi, the physical instability created by the hollow trunk, the wet soil or the burden of ice, or the breeze that finally brought it crashing to the ground. There is even ambiguity in the cause of a massive blowdown: whereas the wind is an obvious cause, the fact remains that the stand would not have blown down if it had been young rather than mature, i.e. age is also a 'cause'. Thus, each disturbance has a chain or network of causation, and the kind, degree and scale of its impacts are determined by a multiplicity of factors. Put another way, a moderate gust of 50 km $h^{-1}$ will disturb one stand but leave another unaffected, depending on the state of the stand and its circumstances.

Despite these uncertainties, the storm that blows a whole forest down overnight is obviously a disturbance, or is it? From the point of view of an individual person living near that forest there is little doubt that this is the event of a lifetime, likely to cause as much disturbance to that person's peace of mind as it does to the forest's structure.

However, storms recur and the forest will again grow to maturity, but in 200-300 years time the forest may again be blown flat. Since 200-300 years is well within the natural span of many forest dominants, it can be argued that blowdowns are a continuing factor in the life of the forest and the evolution of the species it contains. Events that count as disturbances on one time-scale may be ongoing ecological factors on another.

These points are made not to confuse but to emphasize that ecological disturbances are rarely, if ever, discrete events with sharp boundaries and an obvious single cause. Rather, they are often ill-defined in space and time, and the proximal cause may be insignificant in comparison with the multitude of contributory factors.

Pickett and White (1985) have provided a definition of disturbance that incidentally conveys some of the uncertainty: 'a relatively discrete event in time that disrupts ecosystem, community or population structure and changes resources, substrate availability, or the physical environment'. Attempts have been made to distinguish catastrophic disturbances from ordinary disturbances; indeed there is a clear difference between a stand-destroying storm and a wind that snaps a few branches, but the difference is just a matter of degree. Whether one is considering a single kind of disturbance or the cumulative load of disturbances sustained by a single forest, we are dealing with a continuum.

## SOURCES OF DISTURBANCE

Several natural agencies disturb forests and this section briefly describes their immediate impacts and interactions.

## Wind

Wind disturbs a very wide range of forest types (e.g. Foster & Boose 1995; Pontailler *et al.* 1997) (Fig. 4.1). It breaks branches and trunks and it blows trees over. A falling tree or branch usually falls to the ground, although a small proportion comes to rest leaning on other trees. A falling tree or branch commonly hits other trees and shrubs, crushing some, stripping branches off others and knocking over a few. Large trees high on steeply sloping ground may skittle numerous trees standing immediately below them, thereby generating distinctive elongated 'avalanche' gaps.

Fallen trees and branches tend to be clustered. It is common for groups of trees to fall together and in the same direction (e.g. Bouchon *et al.* 1973). Even where single trees fall, the branchwood is clustered where the crown fell, leaving the ground clean where the tree originally grew. Fallen trunks of narrow-crowned trees occasionally embed themselves into the ground with the impact of the fall, but generally trunks lie close to the ground surface or form a low bridge supported by major branches and rootplates.

Trees that snap leave a snag. Those that break near the base leave only a ragged stump, but breakage may occur higher up the trunk, often just below the main crown branches. These higher breakages leave a snag, which usually dies within a few years, often after sprouting weakly from the shattered top. As they decay, snags lose their bark and eventually break up. The roots, which remain moist in the soil, decay faster than trunks, so snags usually fall when partly decayed.

Blown trees form an upraised rootplate, supported by roots on the downwind side, many of which remain unbroken and functional. Root-plates bring a ball of earth and rock with them as they fall, creating a pit on the upwind side.

Trees that have been tipped over in this way often survive the fall, although the chance of this happening depends on the species and circumstances. Thus, a *Tilia* that falls into a gap will generally sprout vigorously from both the rootplate and the prostrate trunk, while a *Fraxinus* which falls into the shade of a neighbouring tree will probably die.

Wind is capable of completely levelling a stand, leaving little more than a few underwood shrubs, saplings and a few snags standing. Disturbance on this scale is achieved by tornados and hurricanes and by the stronger gusts of storms as they hit exposed slopes. Landform can concentrate wind into gulleys and can create eddies that devastate stands beyond a ridge. Generally, however, storm disturbance is patchy, leaving groups of fallen trees within a matrix of unscathed stands, or groups of standing trees in otherwise levelled forests.

## Fire

Fires are the most easily appreciated type of natural disturbance and the one most readily controlled by foresters (Heinselman 1973) (Fig. 4.2). Under natural conditions they start with lightning strikes, which smoulder in the litter and undergrowth. They become stand-destroying infernos when fires run into stands with large fuel loads and a multilayered structure, which provides a ladder up which the fire can climb into the canopy. Once a ground fire becomes a canopy fire it is maintained more effectively by convection and spreads faster by wind and sparks. When well established in inflammable coniferous forest with the wind behind it, a fire can burn hundreds of thousands of hectares (Pyne 1982).

Ground fires burn litter, ground vegetation and saplings but leave large trees intact. Two trees standing close together act as mutual heat reflectors, so are more likely to be killed; moderate fires thus tend to leave a well-spaced stand of larger trees. Crown fires appear devastating and leave groves of blackened snags, although the capricious pattern of fires generally allows groups of trees to survive. Furthermore, heat is directed upwards, so that buried seed, rootstocks, patches of ground vegetation and small mires survive the severest conflagrations.

## Drought

Unlike fires and storms, which occur on a particular day, droughts develop over a season or two (Hursh & Haasis 1931; Clinton *et al.* 1993). Initially trees are merely stressed and growth slows, but as the drought endures growth ceases and leaves fall prematurely. However, stress induced by heat and water shortage may occur regularly within the normal year-on-year variation and cannot be counted as a disturbance. Droughts disturb forests when they kill trees or damage them so severely that they decline and die prematurely.

Trees killed by severe droughts gradually disintegrate from the top. Twigs, small crown branches and, eventually, major branches decay and fall, until the roots rot enough for the whole snag to fall. Trees damaged by drought, and the associated excessive dry heat, lose most of their crown twigs and may effectively stop growing for several years, although vigour may be restored eventually by the growth of epicormic shoots from the trunk and crown branches. The rejuvenating crown may contain several dead crown branches for a decade or more after the drought. Strips of bark may also be killed during a drought and this admits pathogenic fungi to hitherto sound trunks, leading eventually to premature death.

## Biotic

The interactions between plants, animals and trees are sometimes substantial and sharp enough to cross the threshold into disturbance. Perhaps the most obvious are insect epidemics, which can kill canopy trees over large areas (Blais 1965), and localized population explosions, such as bark beetle in accumulations of dead wood, which go on to kill standing trees. Fungal infections are ever present, but occasionally groups of trees are killed together.

The impact of mammals is perhaps more difficult than most to relate to disturbance. Deer and other herbivores are, or were, a natural component of the forest, although they only become a disturbance when their numbers increase or decrease beyond the range of recent fluctuations. Even when deer are clearly far more numerous than would naturally be the case, high browsing and grazing pressures become more a chronic condition than a discrete disturbance. High browsing pressures are felt most keenly by shrubs and saplings, so that the impact of herbivores on forest structure takes many years to develop. Deer browse selectively and thus influence forest composition (e.g. McInnes *et al.* 1992). If their populations depart from the natural range for no more than a few years, their long-term impact may be barely detectable in the age-class distribution, but where high deer numbers persist they influence structure and composition for decades. For example, in Bialowieza Forest, Poland, heavy browsing between 1892 and 1915 admitted a cohort of *Salix caprea* that lasted for 50-70 years (Falinski 1998) and generated changes in the balances between forest dominants that are still continuing (Bernadzki *et al.* 1998).

Small-scale disturbances are commonplace. Wild pigs and cattle disturb the soil surface and create wallows (Falinski 1986). Woodpeckers excavate holes, which may weaken trunks. Beavers fell small trees and dam streams, creating pools and killing trees whose roots become submerged. Eventually, the dam breaks and the resulting moist glade becomes a preferred feeding ground for herbivores. Vole populations may explode for one summer, during which many saplings may be killed by ring-barking at the base. All these effects can be regarded as disturbances on a small scale in time and place but remain ordinary conditions of the forest when regarded on larger scales.

### Water

Heavy rains and floods rarely disturb forests, even those growing on floodplains. Cloudbursts occasionally cause mudflows on slopes and spates in small streams, both of which will uproot trees (Hack & Goodlett 1960). Where the currents flow strongly, water can bend and break shrubs and smaller trees. Ice and floating woody debris will abrade trunks, but rarely to the point of damage.

The main disturbances due to water are long term. Channel movement within floodplains uproots forests at one point but deposits the alluvium elsewhere and thus provides a substrate that is colonized by woodland. Channel movement takes many forms but the general pattern is of elongated patches of bottomland forest being destroyed on the outside of meanders, while elongated shoals are simultaneously built up on the inside.

### Ice

Ice and snow can break branches and may bend saplings and small trees to the ground. Lake ice will generate a disturbed zone round islands and river-borne ice will abrade floodplain trees. In subalpine and other exposed forests, where icy winds come mainly from one direction, the ice causes 'flagging', i.e. killing twigs on the upwind side. Accumulated snow generates avalanches, which can sweep forests from the mountainside, although most avalanches follow tracks which are so predictable that avalanche-proof trees have evolved to grow on them (see also Volume 2, Chapter 3). Thus, in most circumstances these kinds of impact form a normal part of the forest.

Ice and snow become disturbing at a local scale when, for example, an unusually heavy glaze storm breaks numerous crown branches (De Steven *et al.* 1991). It becomes an event, barely a disturbance, when a late spring snowfall flattens saplings and weak basal shoots. Since some species will be killed by this whereas others will propagate themselves by layering, such an event can influence forest composition.

### Topography and landform

The physical environment can itself generate disturbances to forests. The most spectacular, if rare, are earthquakes, volcanic eruptions and meteorite impacts, which can devastate forests over large areas. Landslips can be equally devastating on a local scale but are also rare. Rockfalls from cliffs may generate avalanche-type gaps in the upper margins of the forests below.

## DISTURBANCE REGIMES

Disturbance events and agents of disturbance can be described in isolation but they rarely occur like that in practice. Rather, they interact and combine to form disturbance regimes and thus ensure that most changes in forest structure and composition have a chain of causation: one thing leads to another (e.g. Foster 1988).

Numerous interactions between disturbances have been recognized. For example, a blowdown or a pest outbreak, which generates substantial concentrations of dead wood, initiates circumstances in which a fire is more likely. Wind may break only those trees that have already been weakened by fungal infection, itself originating in fire scars (Matlack *et al.* 1993). Fungi may infect only those trees that have been damaged by drought, or when bark has been stripped by squirrels or deer. The widespread fire of 1644 in the Allegheny Mountains, Pennsylvania, was probably associated with a drought (Hough & Forbes 1943).

Vulnerability to particular disturbances changes with the state of growth (White 1987). If a forest has been free of disturbances for many years, it is likely to have a large stock of mature and weakened trees that will be particularly vulnerable to the first storm, fire or drought to materialize. Conversely, the second storm in 3 years will have little impact if all the vulnerable trees were brought down by the first storm. A particular storm may devastate a mature stand, while leaving an adjacent young stand of the same species barely ruffled. Fires are most likely to destroy young and old-growth stands, where fuel loads are high, but are less likely to do so in middle-aged stands. Each site has its unique history of disturbance, and the impact of any particular disturbance is a function of that history.

The distribution of disturbances is uneven, or patchy, at all scales from single trees up to whole countries or states (e.g. Canham & Loucks 1984). This was exemplified by the storm on 16 October 1987 in south-east England, which brought down 15% of the standing timber volume over several hundred square kilometres (Peterken 1996). Although initial perceptions were of all-embracing devastation, it quickly became obvious that some districts escaped with little damage, whereas the woods in others had been levelled, i.e. the storm appeared to be 'streaky' at a scale of several kilometres. At a much smaller scale, windthrow patches within individual woods were generally small, but the pattern was largely inexplicable. Whereas in some sites blow-downs were associated with wet ground or exposed slopes and gulleys, most windthrown groups appeared to have been no different from nearby groups that remained standing. Patchiness was evident at all scales from districts down to parts of individual woods, and at the smallest scale there appeared to be a strong element of chance in the pattern of gaps.

Despite the random element, disturbance regimes vary in a more-or-less predictable way within a catchment. Forests on the floodplain are disturbed mainly by channel movement, wind-throw and perhaps the effects of prolonged unsea-sonal inundations. Drought is likely to affect stands on upper slopes and plateaus, but leaves stands on flushed lower slopes and the deeper soils of the river terraces untouched. Wind impacts will be determined partly by topography, particularly in regions with a marked prevailing wind direction: slopes facing the wind and stands on the plateau will be particularly vulnerable, and perhaps also stands just over the brow of a ridge, which can be brought down by eddies in the flow.

Likewise, fires tend to brake on ridges and stop at lake margins, so that the pattern of the fire will be partly determined by the landform and the lake pattern. Rockfalls and landslips will be confined to the steepest ground. Grazing may be concentrated in places where the vigour of tree growth is reduced over thin or permanently wet soils.

Taken as a whole, therefore, a catchment constitutes a mosaic of different disturbance regimes, which are likely to be manifested in a mosaic of different stand structures and composition. Equivalent variation occurs at a continental scale. Thus variation in climate results in variation in disturbance regimes and thereby partly determines broad-scale forest patterns.

Disturbances also vary in intensity, although to some extent this is a matter of scale. Thus, a low-intensity storm may blow down no more than 0.1% of the canopy trees in a forest, but in the gaps actually created the disturbance is no less intensive than it would have been in a greater storm. Nevertheless, at a particular scale there is real variation in the intensity of many types of disturbance. For example, a single fire can consume all the trees in a stand; or it may merely singe the underwood as it smoulders through dry litter; or it may burn the small trees and saplings, leaving dominant trees as a form of shelterwood. Similar variation in intensity can be expressed by droughts, pest outbreaks and periods of ice damage. It has become customary to distinguish between catastrophic (i.e. stand-destroying) and non-catastrophic disturbances, even though these are no more than the ends of a continuous variable.

Disturbance regimes are thus the characteristic mix of disturbances experienced by a forest. At the smallest scale, each point in a forest will have experienced a unique sequence of disturbances that go some way to explain its current composition and structure. At larger scales, disturbance regimes can be expressed in two ways.

1. Qualitatively, by listing the kinds of disturbance encountered and describing their frequency, interactions and patterns. This may be all that can be achieved in forests subjected to a variety of disturbances, none of which is dominant.
2. Quantitatively, by calculating various parameters, such as frequency, return time and time to disturb the whole of the forest. This is realistic and appropriate where one form of disturbance is dominant. It works best in fire-dominated forest types, where stand-replacing fires leave a well-defined patchwork of even-aged stands, on whose older trees the scars of past fires can be dated (Heinselman 1973; Zackrisson 1977).

Forest disturbances take two main forms: gap formation and stand destruction. The former occurs when single trees or small groups of canopy trees die or fall, leaving a scatter of gaps in a matrix of intact canopy. The latter occurs when all canopy trees are killed or blown down. Gap formation is a chronic process characteristic of most temperate forests in most decades (Runkle 1982) and is commonly caused by the interaction of wind and biotic disturbances. Stand destruction is an event that is characteristic of forests in some extreme environments (e.g. Sprugel 1976), of seasonally dry climates and in hurricane belts. Fire is the most common agent of stand destruction, although wind and invertebrate pests can also destroy stands on a large scale. The history of many forests can be summarized as long periods of gap creation, punctuated at intervals of centuries by stand destruction and subsequent regrowth (Henry & Swan 1974; Stewart 1988). Intermediate scales and intensities of disturbance are frequent. For example, extensive fires may leave a shelterwood of large trees, or droughts may weaken or remove a common canopy species; both may leave stands in which 'gaps' can be said to form the matrix for the remaining trees. Rarely, some forests appear to escape disturbance, for example the montane bristlecone pine park-lands, where the vegetation is too sparse to allow fires to spread.

## TREE AND SHRUB RESPONSES

The various responses of trees and shrubs to individual disturbances combine with the overall disturbance regime to determine forest composition. Each species posesses a unique combination of responses that enable it to survive and reproduce. The key variables between species relate to the form of reproduction, dispersal, tolerance of shade, growth rate and longevity.

Species vary in the degree to which their fruits or other propagules are dispersed. Whilst most propagules fall close to parent trees, a few species achieve long-distance dispersal. The lightweight fruits of *Betula, Populus* and *Salix* are widely dispersed by wind. These and other species growing in regions with persistent winter snow are widely dispersed as fruits are blown across the icy surface. Species with fleshy fruits are dispersed by birds, either passing through the intestinal tract (e.g. *Sorbus, Malus)* or being transported as food (acorns of *Quercus).* Species that grow close to rivers are dispersed downstream if their propagules fall in the water. Efficient dispersal enables species to occupy disturbed ground rapidly and facilitates speedy long-distance movement of populations. Species with a limited dispersal ability must have a capacity to survive on-site, usually by vegetative reproduction and/or living to great ages.

Species that demand high light intensities must be able to occupy disturbed ground rapidly and maintain themselves in the upper stratum. Such shade-intolerant species tend to be either shortlived shrubs or fast-growing trees with strong leaders and widely dispersed foliage.

Tolerant species survive in shade but their growth is slow and most grow faster in higher light levels. Their shade tolerance enables them to survive in undisturbed stands indefinitely and to enter established stands, where they commonly form a persistent underwood of shrubs, saplings and subcanopy trees. Shade tolerance is a continuous variable (Lorimer 1983) so that intermediate responses are widespread. Furthermore, it can vary with age, such that seedlings are shade tolerant but saplings of the same species become intolerant as they age (e.g. *Fraxinus excelsior).*

Another key variable is growth rate, specifically growth in height. Fast-growing species will maintain their position in the canopy, whereas slow-growing species remain in subordinate strata and must therefore bear shade in order to survive.

These three variables are correlated enough to allow a useful distinction to be drawn between two survival strategies.

*Strategy A:* widely dispersed, shade-intolerant species with rapid height growth. These species are equipped to respond rapidly to disturbance. They colonize disturbed ground and gaps in established forest and maintain their position in the canopy. *Strategy B:* shade-tolerant species, capable of surviving with slow growth rates and commonly limited in their ability to disperse. They are equipped to survive on the sites they currently occupy. Whilst they are commonly found in the underwood and subcanopy below a canopy of strategy A species, they are usually capable of persisting in the canopy. Other important variables can be related to these strategies, though there seems to be no close connection between strategy and longevity. An ability to live to a great age prolongs canopy residence time and can thus form an advantage for strategy A species, several living to 300 years or more (e.g. *Pseudotsuga menziesii, Picea sitchen-sis, Quercus robur, Liriodendron tulipifera, Pinus sylvestris)* and forming some of the most enduring and impressive forest stands. Great longevity is also advantageous for tolerant species, enabling them to persist long enough in the underwood to grow into large canopy trees (e.g. *Tilia cordata)* or massive subcanopy individuals (e.g. *Ilex aquifolium, Taxus baccata),* though in most cases these species probably live longer if they occupy the canopy or prevent inherently taller species growing through them. On the other hand, many strategy A species are short-lived, lasting no more than a century (e.g. *Betula pubescens, Populus tremula, Salix caprea)* or two centuries *(Alnus glutinosa, Fraxinus excelsior, Populus nigra),* even when they succeed in staying in the upper strata. Furthermore, strategy B species are not necessarily long-lived: beech and Norway spruce may take well over a century to reach the canopy, but then have only 100 years of residence once they get there.

Early, large and regular seed production is characteristic of some short-lived strategy A species (e.g. *Betula pendula),* which facilitates colonization whenever and wherever opportunities are created by disturbance. Otherwise there seems to be little correlation between seed production and strategies. Late, irregular seed production is characteristic of mast trees, such as *Quercus robur* and *Fagus sylvatica,* and is said to facilitate regeneration by periodically outproducing the capacity of predators.

A capacity to reproduce vegetatively is also common in both strategies. Most broadleaved species will sprout from stumps after the main stem has been destroyed or knocked down in a disturbance, and some species (e.g. *Tilia ameri-cana)* regularly reproduce in this way even in undisturbed stands.

This is a classic strategy B device, which effectively restocks storm-damaged forests (Peterson & Pickett 1991) and indefinitely prolongs occupation of the site. On the other hand, several strategy A species do not sprout strongly, for example *Betula pendula.* However, the correlation with strategies is limited and there are many exceptions. For example, floodplain willows (e.g. *Salix alba)* will root from almost any fragment of branch, which enables them to take advantage of severe disturbances. Aspens *(Populus tremula, Populus tremu-loides)* grow short-lived trees from persistent rootstocks, thereby combining the two strategies. *Fagus grandifolia,* a classic tolerant species, generates suckers from a persistent rootstock.

Germination rates also vary significantly. Fast germination enables species to take immediate advantage of temporarily favourable conditions. When the germination of a particular seed crop is spread over a number of years, the chance of germinating in favourable conditions is increased at the cost of a reduction in the number of ger-minants at any given time. Prolonged dormancy appears to be associated more with light-demanding species, for example *Cytisus scopar-ius* can persist in the seed bank until gaps appear. Perhaps the most extreme adaptation is serotiny, in which a proportion of the seed is retained on the tree until released by a fire into a suitable seed bed and freedom from competition (e.g. *Pinus banksiana).*

There are thus many combinations of characteristics and responses that enable species to survive in forests where disturbance is irregular. In undisturbed conditions stragegy B species predominate, leaving strategy A species to a fugitive existence in gaps and localized disruptions, for example in southern Britain *Tilia* spp. were dominant until widespread prehistoric forest clearance (Greig 1982). After prolonged quiet, a massive disturbance gives strategy A species an opportunity to thrive, provided they can react faster than strategy B species regenerating vegetatively. In environments that are frequently and substantially disturbed, strategy A species tend to predominate (see example of ice-dominated forests below).

## FOREST STRUCTURE AND PATTERNS

The interactions between disturbances and forest trees generate a complex range of forest structures and patterns (e.g. Elliott-Fisk *et al.* in Barbour & Billings 2000). Perhaps the best route through this complexity starts with two contrasting stand structures (for descriptive purposes a stand is taken to be a compact patch of woodland of about 2 ha).

1. *Even-aged:* stands are even-aged if the canopy trees all started growth in approximately the same year. A limited age spread of perhaps 10 years is permissible within the definition. Such stands originate after a stand-destroying disturbance or an abrupt change in land use. Colonizing trees may take longer than 10 years to fill the available space, but as the stand develops its age range narrows: latecomers will have smaller, lower crowns than those that started growth immediately and this disadvantage ensures that their mortality rate is greater.
2. *Mixed-age:* stands comprising a wide range of age-classes, from old trees to saplings, intimately intermixed. Such stands tend to form a mosaic of even-aged groups at the scale of one or two canopy trees. They function by a process of relatively constant gap formation, in which gaps formed by the loss of canopy trees or major branches are filled by groups of saplings. These groups compete amongst themselves as they grow, so that eventually only one or two trees become established in the canopy, thereby achieving a 1:1 replacement of the original gap-forming tree. Such stands can only survive where there has been

no catastrophic disturbance within the adult lifetime of the oldest canopy trees.

Even-aged stands develop through a sequence of stages, which are explained in Chapter 9. In effect, they become steadily more mixed-age as they develop, initially through recruitment in the underwood and eventually through mortality of canopy trees and replacement with individuals initiated later (e.g. Aplet *et al.* 1988). In the absence of catastrophic disturbance, an even-aged stand eventually becomes mixed-aged, a stage described as 'shifting-mosaic steady-state' by Bormann and Likens (1981) or 'old-growth' by Oliver and Larsen (1996)

This relatively simple model is complicated by non-catastrophic disturbances. Gaps created by storms, for example, are often irregularly patchy. Canopy disturbance created by droughts may take the form of an irregular thinning. Likewise, fires of intermediate intensity commonly leave a fairly even scatter of large trees. Regeneration follows both patterns of disturbance, so that moderate disturbances eventually generate both small-scale patches of younger growth within a hitherto even-aged stand and two-storied stands in which the survivors from the previous generation persist in a matrix of younger growth.

Partial disturbances give rise to stands intermediate between the mixed-aged and even-aged stands defined above (e.g. Maissurow 1941). They are mixed in the sense that different age-classes are mixed together at a small scale, even at canopy level, but contain distinct generations, i.e. concentrations of individuals within narrow age bands. Jones (1945) recognized that strong even-aged elements within mixed-age stands were common within virgin temperate forests.

At the stand scale, a plot of the age (and often also the size) of trees demonstrates pulses of recruitment, and it may be possible to partition the stand into two to four more-or-less distinct cohorts. In truly mixed-age stands, the age and size distributions take a negative exponential form (Hough 1932). In stands with strong even-aged elements, the relationship remains a negative exponential but with peaks and dips. Thus, for example, the size distribution of *Quercus mongolica* in virgin old-growth in Hokkaido showed a fairly smooth, continuous, J-shaped distribution, even though the population comprised five distinct generations spanning 500 years (Sano 1997).

A second complication arises from the tendency of gaps to enlarge once they have formed. For example, windthrow gaps within coniferous forests become foci for populations of the beetle *Ips typographies*, which then colonize and kill trees standing by the gaps. In forests of all kinds, once a gap has been opened by wind, neighbouring trees are more exposed to subsequent winds and thus more likely to be blown down (Runkle & Yetter 1987). As gaps enlarge, so they tend to coalesce, until the stand ceases to be a canopy punctuated by discrete gaps and becomes an irregular scatter of fully grown trees in a matrix of shorter, and usually younger, individuals.

European forest ecologists, led by Liebundgut (1959), have not recognized gaps as distinct features but have attempted to represent the structure of old-growth stands as a patchwork of different stages of stand development. Each stage is defined as a distinct spatial arrangement of trees of different sizes and ages and is represented as a developmental link between two or more other stages, the whole representing a cycle of stand development with several pathways (e.g. Zukrigl *et al.* 1963; Mueller-Dombois 1987). Several different versions of the classification of stages and the dynamic cycle have been published. Maps have been prepared to show the patchwork of stages in particular forests (Mayer *et al.* 1980).

In practice, neither gaps nor the Liebundgut cycle fully represents the complex structure and dynamics of undisturbed old-growth. Gaps are readily defined and delimited in the early stages of canopy break-up but become far less distinct as break-up progresses and the stand becomes multi-layered. Furthermore, most gaps are small, so that the question of what counts as a gap is crucial in any quantification, especially in stands where the crowns of canopy trees rarely touch, let alone overlap, for example *Pseudotsuga menziesii* old-growth. In the case of the Liebundgut stages, the patches so clearly defined and mapped on paper are found on the ground to be ill-defined and poorly delimited. Intermediate conditions are common. Each patch of a particular stage is seen to be internally heterogeneous, i.e. to be a micropatchwork of different structures. The stages and maps based on them represent reality as perceived at a particular scale: both the classification of stages and the maps would have been quite different if a smaller scale had been adopted.

In so far as they can be delimited, gaps have internal structure. The distribution of dead wood is generally irregular, for example a gap created by the fall of a single tree has an accumulation of crown branches and debris on one margin, leaving the rest of the gap with a clean floor. In many forest types the underwood of shrubs and groups of advance regeneration survives gap formation: its original irregularities will be increased by some flattening as the canopy tree falls.

In larger gaps it is common for one or two subcanopy trees to survive, thereby retaining a light shelterwood within the gap. Irrespective of size, gaps have an internal geometry which ensures that in the Northern Hemisphere the northern side receives more direct sunlight than the southern side and that the eastern side becomes drier in afternoon sun than the western side, which receives direct sunlight during the more-humid mornings. Furthermore, light intensity is greater at the centre of gaps than at the edges. In general, larger gaps permit a larger representation of shade-intolerant species (Poulson & Platt 1989). Internal heterogeneity created by fallen trunks and branchwood generates irregularities in browsing intensity and small-scale shading patterns, and thus also contributes to the variety of regenerants (Hutnik 1952).

Both gaps and the representation of stages have been used to quantify canopy turnover. Gaps are formed, or canopy space is vacated, at about 1% per annum. The rate varies from year to year (Runkle 1982) and on a scale of decades (e.g. Poulson & Platt 1996) but where it has been measured in different forests, gaps have formed at 0.5-2% annually. This implies a residence time of about 100 years for trees that reach the canopy. Where the extent and duration of growth stages have been quantified, it has been possible to compute the duration of the growth cycle at some 300 years and to determine the degree to which the age distribution of patches departs from the theoretical steady state (Mayer & Neumann 1981).

In practice, individual trees live far longer than 100 years. Firstly, trees take several decades to reach the canopy, particularly those shade-tolerant species that remain virtually static when they are heavily shaded. Even fast-growing light-demanding species take 30 years or so to reach canopy height. Secondly, some canopy species are inherently short-lived, leaving space for other species to reside longer in the canopy. Thirdly, gap formation is not absolutely correlated with age of trees, so that some individuals live far longer than the majority. In temperate forests, individuals of oak, tulip, hemlock and lime regularly achieve ages of 300-500 years or more (Lorimer 1980; Runkle 1981).

When forest structure is considered at a landscape scale we see that the fundamental difference between even-aged and mixed-age stands is a matter of scale. A forested landscape made up entirely of mixed-age stands has the same size distribution and age-class distribution as a landscape covered in even-aged stands, each of a different age. It is only when the forest is considered at a stand scale that even-agedness is perceived. A landscape of even-aged stands is usually mixed-aged in aggregate, i.e. the various stands comprising the landscape represent a range of ages (e.g. Heinselman 1973). Thus, at a very large scale all forested landscapes are mixed-aged, although the scale of their even-agedness varies from groups no larger than a single canopy tree up to patches covering several thousands of hectares. Some historic fires have burned over whole counties, leaving patches of hundreds of thousands of hectares that would have regenerated as a massive even-aged stand, for example the widespread 500-year age-class of *Pseudotsuga menziesii* in the Cascades (Franklin & Hemstrom 1981).

Variation in the scale of the patchwork has several consequences for forest composition and structure. A very small-scale patchwork implies a fairly steady rate of gap creation when computed at a large scale, which in turn implies that forest structure will remain more or less constant. A large-scale patchwork implies that at moderate scales (say individual catchments of first- to fourth-order streams) forest structure can change widely with time according to the irregularities of disturbances, for example Yellowstone (Romme 1982). Frelich and Lorimer (1991) used growth stages and transition times between each stage to compute the state of the landscape in a region of moderate disturbance, and found that most of the forest remained old-growth for most of the time.

## DEAD WOOD

Dead wood, or coarse woody debris (CWD), is an integral component of the forest (Maser *et al.* 1988). It is generated not only when trees are killed by disturbances but also when trees or individual branches die in the ordinary course of competition between individuals and change in stand structure. It also develops within the heartwood of living trees. Above-ground CWD thus takes the form of fallen wood, dead standing stems (snags and stumps), dead branches on living trees and decay columns within living trees.

The amount, distribution and rate of decay of CWD in natural temperate forests varies widely (Harmon *et al.* 1986). CWD generally increases with stand age but is also high in young-growth stands growing through the debris left by a storm or other disturbance (e.g. Spies *et al.* 1988).

The amount of CWD varies through time in any particular stand: pulses of CWD input caused by disturbances are superimposed on a steady background input from stem exclusion processes. CWD distribution is also patchy, i.e. it is concentrated in recently disturbed forest, which is itself patchy, and accumulates in gulleys and debris dams in rivers.

The quantity of CWD is often measured as the volume of snags and fallen logs. In most old-growth temperate deciduous stands the values generally fall between 50 and 140 $m^3 ha^{-1}$ (Kirby *et al.* 1998), to which should be added perhaps 20-30$m^3ha^{-1}$ for snags.

Values from other forms of temperate forest tend to be much higher and comparable with volumes in boreal forests. Thus values in the *Picea sitchensis-Pseudotsuga menziesii* forests of the Pacific North-west have been estimated as ranging from 600 to over 1400 $m^3 ha^{-1}$ (Harmon *et al.* 1986), and for *Nothofagus* forests in New Zealand as 370-800$m^3ha^{-1}$ (Stewart & Burrows 1994).

## DYNAMIC PATTERN OF FOREST COMPOSITION

Interactions between tree species and between them and disturbance regimes generate distinctive assemblages of species, forest dynamics and landscape patchworks (White 1987). These are best illustrated by examples at the ends of the range of disturbance frequency.

### Highly disturbed environments

Where disturbances are frequent, large scale or severe, forests tend to be dominated by intolerant and fast-colonizing species. Three examples illustrate the range of circumstances.

#### *Floodplain forests*

Floodplain forests are influenced by channel movement, which destroys mature stands but reworks the deposits into new shoals. In northern temperate regions, these are colonized mainly by *Salix*, *Alnus* and *Populus* species, which grow into even-aged, often monospecific stands. The sequence of channel movements is manifested as a pattern of elongated even-aged stands, whose age increases with distance from the channel. Tolerant species colonize beneath these pioneer stands and, given sufficient time, develop into mixed old-growth. In northern temperate deciduous regions the principal long-term dominants are *Ulmus*, *Fraxinus*, *Quercus*, *Carya*, *Sassafras* and, in the Pacific North-west, *Picea sitchensis* and *Pseudotsuga menziesii*. At any one time, the pioneer stands tend to predominate near the present channel, and the mixed old-growth tends to survive in elongated patches at some distance from the river (Hupp 1988; Cordes *et al.* 1997).

#### *Ice-dominated forests*

The classic type is the wave-regenerated *Abies* forests of the eastern USA and Japan (Sprugel 1976). Exposed mature forest degenerates when foliage is stripped by ice and wind from trees that have already lost the vitality of youth. Death of exposed trees exposes others, leading to a 'wave' of mortality, which moves steadily through the forest in the direction of the prevailing wind. Regeneration starts within the degenerating stands and grows vigorously in the lee of slightly older stands. The forest as a whole takes the form of a series of parallel waves, which move through the forest at 1-3 m annually on a return time of 60-70 years. This perpetual recycling ensures that *Abies balsamea* remains dominant and that the longer-lived *Picea rubens* is perpetually excluded.

*Fire-dominated forests*

The most widespread form of highly disturbed forest is dominated by fire (e.g. Heinselman 1973; Romme 1982). Most boreal forests are naturally fire-dominated, but so too are Mediterranean forests and the forests that fringe extensive grassland and desert regions. Each region has a suite of species that undergo a characteristic succession after fire, for example in Scandinavia *Betula-Pinus* develops into *Pinus* dominance, which is then succeeded by *Picea.*

Exceptionally, some patches remain unburned, which allows the pioneers to be completely displaced, although generally fire returns in good time to ensure that pioneer species remain a permanent feature of the forest. In fact, there is much variation in return time, associated with variation in topography, ground vegetation and the configuration of water bodies (Zackrisson 1977). Nevertheless, fires were frequent enough to maintain most boreal forests as young or maturing stands, not old-growth. Similar fire-dominated regimes control other forests, such as the *Eucalyptus* forests of Australia and the *Pinus*-dominated forests of the coastal plain of the south-eastern USA (Chris-tensen 2000). However, in western North America many tree species not only withstand fires once they have achieved a moderate size but also grow to great size and age, thus generating the monumental forests of *Sequoiadendron giganteum, Sequoia sempervirens* and *Pseudot-suga menziesii* in which old individuals may bear the scars of several fires. On the margins, the boundary between mesic- and fire-dominated forests advances and retreats according to the history of fires.

## Relatively undisturbed environments

At the other end of the range are mesic forests growing in relatively undisturbed environments. These are not disturbance-free; instead, catastrophic disturbances are rare enough to allow most of the forest to develop into old-growth, where disturbances are small-scale events.

Over much of the north temperate zone, a distinction can be drawn between *Fagus* forests and mixed deciduous forests (Ellenberg 1988; Barnes 1991; Jahn 1991). *Fagus*-dominated forests are found mainly in northern latitudes and submontane elevations, where *Fagus* spp. often share dominance with conifers, for example *Tsuga canadensis* in the eastern USA, *Abies alba* and *Picea abies* in central Europe. European beech, *Fagus sylvatica,* is not long-lived, but casts dense shade, is capable of regenerating in the small transient gaps generated in beech forests and is almost fireproof, so that it is able to both dominate the site and perpetuate this dominance. However, it is prone to disaster, in the sense that mature stands are vulnerable to drought and high winds. A few intolerant species, such as *Salix caprea,* maintain a foothold in large gaps, while the shade-tolerant *Acer pseudoplatanus,* like *Acer saccharum* in the eastern USA, fills gaps, competes in advance regeneration and can grow into mature stands. Diversity is maintained partly by the tendency of species not to regenerate under themselves. Thus, *Fagus grandifolia* and *Acer saccharum* (Fox 1977) in the USA and *Fagus syl-vatica* and *Abies alba* in Europe have been reported to alternate, thereby maintaining a small-scale mosaic of different dominance, though in the former case coexistence has been ascribed to different responses to light intensity (Poulson & Platt 1996). Occasionally, such stands are destroyed by storms (Peterson *et al.* 1990), whereupon pioneer species dominate the regrowth, although beech thrives in the underwood and eventually restores its position.

Mixed deciduous forests are those which lack the dominating influence of beech. At their greatest development, for example in the southern Appalachians, they comprise a mixture of several dozen species, each with the capacity to occupy the canopy or subcanopy (Lorimer 1980) (see also p. 49). A wide range of genera are represented, notably *Acer, Quercus, Fraxinus, Carya, Aescu-lus, Betula, Castanea, Tilia, Magnolia, Carpinus* and *Halesia,* forming rich mixtures in which no single species becomes absolutely dominant. Such forests are rarely devastated by any single disturbance and for most of the time are renewed by gap-phase regeneration on a small scale. Pioneer species regenerate in the larger gaps and may dominate the canopy after periods of enhanced gap creation. An example is *Lirioden-dron tulipifera,* which not only forms the tallest trees in the Appalachian forests but can also live for 500 years, enabling it to perpetuate itself through long periods lacking disturbance.

Within these complex mixtures, each species has a distinctive pattern of growth, longevity and regeneration. Furthermore, there is a tendency for individuals of one species to be replaced by another species. The forests comprise a small-scale mosaic of groups of different canopy and underwood species, each with its particular successional trend. The trends in one patch are countered by opposite trends in other patches, thereby retaining the mixture. Recruitment of particular species tends to be irregular, depending on particular combinations of mast-years and disturbances. Composition remains fairly constant overall but at a small scale changes perpetually, except where individual trees replace themselves vegetatively, e.g. *Tilia* spp.

Single-species groups are common in mixed forests. These may develop in response to small differences in site conditions and/or the chance coincidence of an episode of gap creation with heavy seed production by a particular species. Simulations have recently shown that neighbourhood effects may also play a part (Frelich *et al.* 1998). Where there is a high probability that canopy trees will be replaced by individuals of the same species, 10 generations is enough for this feedback to generate small-scale single-species patches. The scale of the patches increases substantially where minor environmental differences result in 5% alterations in recruitment probabilities.

## Interactions and intermediate conditions

The forests described above are stereotypes that represent the ends of the range of variation. In practice, many forest types are intermediate in some respect, although their dynamics and composition can be related to the stereotypes. Some examples illustrate the range of possibilities.

- Forests in which partial stand destruction is commonplace can develop into complex mosaics of tolerant and intolerant species (e.g. Stewart 1988). An example is the *Pseudotsuga menziesii* forests of the Pacific North-west, where saplings and underwood are frequently burned but the overstorey is unaffected. Moreover, light fires tend to be patchy, burning some places but leaving other patches untouched. This allows the overstorey of *Pseudotsuga menziesii* to be infiltrated by *Abies amabilis*, which develops as patches in a mosaic with groups of younger *Pseudotsuga menziesii* in larger burned patches.

- Complex interactions can develop between disturbance regimes. For example, in the southeastern USA *Pinus* forests were maintained naturally by frequent fires. However, where stands happen to remain unburned, *Quercus* and other broadleaved species colonize, form an underwood and eventually dominate the canopy (Iwaschkewitsch 1929; White 1987). Since the broadleaved stands are relatively fireproof they tend to remain undisturbed, thereby generating a patchwork of broadleaved and *Pinus*-dominated stands whose pattern is determined by disturbance history. This capacity for two forest types to develop on one site type is inherent in any region where disturbance becomes less likely as succession proceeds.
- The characteristics of *Fagus sylvatica* in central Europe determines much of the forest pattern. Its capacity to coppice enables it to form distinctive low scrub woodlands with *Sorbus aucuparia* near the treeline, where wind and snow maintain a chronically disturbed environment. In this case, the *Fagus sylvatica* is effectively undisturbed. Elsewhere, in steep slopes in gorges, the stands are chronically destabilized by the fall of large trees growing on insecure rootholds, and this breaks the dominance of beech and enables a mixture of tree species to survive, including *Tilia, Ulmus* and *Fraxinus*.

Disturbances have long-term effects on forest composition. For example, drought rarely destroys stands entirely but confers an advantage on species that survive droughts. Ultimately this is expressed as a distinctive assemblage on drought-prone sites, such as outcrops and convex slopes with thin soils. In this instance, disturbance tends to generate adapted forests, which thus reduces the incidence of disturbance. Alternatively, species that depend on disturbance may both invite it and be ready to survive. The best examples are the fire-dependent *Pinus banksiana* and *Pinus contorta* forests, which are inherently inflammable, particularly those that exhibit serotiny, where long-lived cones release their seed only after fire and which falls into a seed bed free of litter and competing vegetation. A more extreme example is seen in the *Pinus mugo* of avalanche tracks, where repeated disturbance has provoked the evolution of a species that can withstand the disturbance.

The multitude of interactions between trees and disturbance regimes is expressed as different assemblages on different site types and as a range of successional states on each site type. Disturbance should be seen as an integral component of the forest type, not an external destructive force (White 1979). Collectively, disturbances enhance regional diversity by creating distinctive forest types, maintaining a range of successional states, and by enhancing the amount of edge in the landscape.

## Influence of people

The relationships described so far have dealt with natural conditions. In practice, most forest types are overwhelmingly influenced by people:

- Secondary successions to forest on land previously cleared are a response to severe disturbance, i.e. the complete loss of tree cover and the alternative use of land for cultivation or pasture. Pioneer trees are almost always intolerants. The pattern of secondary succession depends on seed sources, so that colonization may be extremely patchy. Colonization often takes place in waves, for example when initial colonization by wind-dispersed trees is followed by bird-dispersed trees once perches have been created.
- Coppice systems (Rackham 1980) enable shrubs and intolerant trees to form a higher proportion of a stand than they would naturally maintain. Despite the constant disturbance of felling, individual trees can sprout again indefinitely and thus live well beyond their natural span. In terms of individuals and small-scale patterns, coppices are probably more stable than the natural forests from which they were derived.

- In woodland pastures (Rackham 1980), regeneration is inhibited by grazing and browsing, although established individuals have little competition and can expand to great sizes. Moreover, the constant lopping of branches for fuel and fodder, followed by regrowth from the pollard, enables individual trees to grow to great ages. Here, too, the effect of management disturbance is to generate unnaturally high stability.

Finally, there is the problem of *Quercus petraea* and *Quercus robur* in Europe. These are long-lived intolerant species that have been abundant in the pollen rain over very long periods (Godwin 1975), which implies that the forests were considerably disturbed, albeit at long intervals. They have been maintained at unnaturally high levels by traditional management, which favoured them as timber trees in coppices and wood-pastures (Godwin & Deacon 1974). Under natural conditions it is possible that oaks were maintained by the high level of browsing and grazing imposed by populations of large herbivores, such as deer, horses and cattle. In fact, it has been proposed that natural forest was more like savannah before prehistoric hunters reduced and domesticated the larger herbivores (Wallis de Vries 1995). However, this seems far from proven, although the possibility remains that herbivore populations maintained a controlling presence on the structure and composition of some types of temperate forest, particularly perhaps those on wet ground and heavy fertile soils (where ground vegetation would be vigorous) and steep dry slopes (where tree growth might be poor) (Bradshaw & Mitchell 1999).

## STABILITY, SUCCESSION AND CLIMAX

The concepts upon which forest ecology developed for many decades centred on the notion of forests as stable places (McIntosh 1985). Change was recognized but changes formed part of a succession that would lead eventually to a stable end-point, or climax state, which over much of the globe was a form of forest. As the name implies, climax forests were the peak of development, an ideal in which biomass, biodiversity and productivity were maximized in self-sustaining groves of stupendous arboreal giants. As the theories developed, change was recognized even in climax vegetation, allowing a cycle of structures within the climax state; further, it was accepted that under some circumstances vegetation regressed 'beyond' the climax to a postclimax state, less productive, with lower biomass and generally less impressive than the climax. Disturbances were seen and understood but they were regarded as destructive, not part of the system.

The climax concept originated as a basis for vegetation mapping (Cowles 1899) and was sustained by its simplicity, the idealism of its time and as a target for conservation measures (Raup 1981). However, it has now largely been supplanted by alternative models. 'Stability' can be defined by the capacity to return rapidly to a former state after disturbance: in this sense assemblages of arable weeds are more stable than ancient forests (Horn 1975). Vegetation is seen to develop through defined stages (i.e. succession) to no particular end-point (Bormann & Likens 1979; Oliver 1981). Succession continues predictably only until the next disturbance. The next disturbance may totally destroy the vegetation and thus reinitiate succession, or it may partially destroy vegetation and thus merely deflect succession. The composition of the early colonists, which may be determined largely by chance or transient conditions, can determine the long-term rate and direction of change (Egler 1954). Thus, in each forest region, succession must be seen not as a single highway but more as a network of possible paths along which the vegetation at any one site moves according to its particular sequence of disturbances and responses (Green 1981; Miles 1988). Even in undisturbed old-growth forests, there is no one-to-one relationship between composition and site factors, merely various probabilities that certain species will be present (e.g. McCune & Allen 1985).

Perceptions of change and stability depend on the scale at which they are considered. The natural forests of a catchment may retain the same suite of species for millennia, although the precise pattern of these species within the catchment may change over decades. It may be perpetually disturbed but retain a constant balance between young-growth and old-growth in the catchment as a whole. Even within 'stable' old-growth stands single-tree gaps will form and the replacement trees will probably not be the same species as the fallen tree (Fox 1977; Runkle 1981). Thus, each of these conditions can be regarded as either stable or unstable according to the spatial scale or time-scale considered. The concept of *minimum dynamic area* has been advanced as an expression of 'the smallest area with a natural disturbance regime, which maintains internal recolonization sources and hence minimizes extinction' (Pickett & Thompson 1978), although this does not imply absolute stability in the proportions of tree species. Estimates of such areas for the forests in the eastern USA were $10^2$-$10^4$ha for forests disturbed only by small-scale gap creation and $10^5$ha for forests controlled by fires (Shugart 1984).

At much larger scales of space and time, tree species migrate individualistically in response to long-term, large-scale climatic oscillations (Davis 1981; Huntley & Birks 1983), which shuffle arboreal assemblages and perpetually alter the context of competition within which each species evolves, i.e. trees migrate, spread and retreat but do not migrate as consistent communities. Individual species and assemblages of species at any particular point may never be quite in balance with their current environment (Raup 1957). At any given time, particular species may be advancing into a stand (e.g. *Fagus* in Britain), whereas others may be relicts (e.g. *Tilia* in Britain).

## PEOPLE AS AGENTS OF DISTURBANCE

Much of this chapter is based on natural forests, where people and their influences are insignificant by definition. Such forests still exist, but in the real world people have long been a factor.

### Seminatural disturbances

The varied roles of people in the disturbance regimes of even 'natural' forest reserves may be clarified with an example. After 50 years of detailed observation of Lady Park Wood in the UK, we can list the disturbances to which it has been subject and the way in which these have been influenced by people (Peterken & Mountford 1995). The wood itself is an ancient seminatural broadleaved mixture that was managed as coppice for centuries; however, in the late nineteenth century a decision was taken to promote it by periodic thinning to high forest, and then in 1940-44 part of the wood was virtually clearfelled to supply timber to the war effort. Since 1945 the wood has been allowed to grow without direct human intervention. The recorded disturbances are summarized in Box 4.1. Evidently, the influence of people has been all-pervasive. The impact of drought and many other disturbances has depended on the age and composition of the stands, which have been determined by the history of management. Eventually this will grow out, but even so most of the biotic disturbances will probably remain seminatural in the sense that the actions of people were essential in the chain of causation. Furthermore, events around the reserve, such as felling and variation in deer control, will remain significant within the reserve. The reserve may appear to be natural but its disturbance regime is no more than seminatural.

## Silvicultural systems

Many commentators have pointed out that felling is equivalent to natural disturbance (e.g. White-head 1982), although the analogy should not be pushed too far. Felling followed by timber extraction removes most of the potential volume of dead wood and generates additional disturbance to the microtopography and soil structure. Felling itself is not identical with windthrow: it leaves low flat stumps in the ground and does not generate ragged snags nor rootplates and tip-up mounds. Death standing is not part of forestry.

The structure, pattern and mode of regeneration of most natural forests are broadly equivalent to the high-forest silvicultural systems practised by foresters (Jones 1945). Within this, perhaps the greatest degree of equivalence to a particular silvicultural system is achieved by fire, which can remove timber and potential dead wood almost as effectively as timber harvesting. In boreal forests fires generate either a shelterwood structure where scattered mature trees remain or patchy retentions where groups escape fire. At a landscape scale, fires generate a mosaic of large- and small-scale patches, whose boundaries commonly coincide with ridges (watershed boundaries to Americans), rivers and wetlands, and this corresponds with large-scale forestry operations. Rotations under a natural fire regime are short (Zackrisson 1977), which is not so different from modern timber production. Prescribed burning after felling and extraction creates ground conditions that resemble those created by natural fires. Virtually all regeneration is from seed. Natural fire-dominated disturbance regimes can readily be mimicked by commercial forestry as either irregular even-aged patches or two-storeyed high forest (Angelstam 1997).

In the wind-dominated disturbance regimes of temperate deciduous forests, the silvicultural equivalents are selection and group-selection systems combined with occasional large coupes containing scattered shelterwood retentions. A reasonably exact mimicry of natural systems would be achieved with minimal thinning, rotations allowing canopy residence times of 100 years, and indefinite retention of groups of canopy dominants. In floodplain forests, where the principle disturbance is channel movement, the equivalent silvicultural system would be long-rotation even-aged high forest.

Traditional coppice and wood-pasture systems have very few natural equivalents. Volumes of dead wood are very small in coppices, less than 10% of natural levels (Kirby *et al.* 1998), or are concentrated in the decay columns of parkland pollards. Forms of coppice can be generated by repeated leader mortality in extreme environments, such as the exposed ridges in subalpine scrub and the broadleaved underwood in fire-dominated regions (New Mexico). Wood-pasture structures can also be generated in the marginal zones between forest and grassland, notably in central California and the prairie margins. Wood-meadow structures can also be envisaged as a natural feature of the stable portions of flood-plains, where luxuriant herbaceous vegetation would inhibit tree regeneration and attract herds of deer and cattle: any tree that does become established would grow rapidly and expansively. Natural circumstances of this kind have vanished from temperate floodplains, but they might have been common in the Pleistocene era of megaherbivores.

## Traditional cultures as part of nature

Whereas we readily accept that present-day disturbance regimes and forests owe their character to both people and nature in varying degrees, we have assumed that there was a time when people had little or no effect. In northern temperate Europe this era closed about 5000 years ago, when expanding agriculture initiated widespread forest clearance. The forests that developed before this time are termed 'original-natural', 'primeval' or 'wildwood'. A few survived in remote or protected areas into historic times. In North America, the original-natural forest has been thought to be more within historic reach. In many places, old-growth forests that have not been directly altered since European settlement survive in the modern landscape, and great swathes of second-growth forest are returning to a condition similar to that which obtained before the original forests were logged. It was against this background that the concept developed of a stable climatic climax as the destination of natural succession.

Latterly, however, the profound role of native Americans in shaping the forests has increasingly been recognized (Denevan 1992). Apart from areas of cultivation and controlled pasture, vast tracts were burned sufficiently to maintain grassland where trees would grow and maintain open forests that might otherwise be closed. Much of the 'natural' forests or 'pristine wilderness' encountered by Europeans was regrowth from 1500 to 1750 after the populations of native Americans had been greatly reduced by European diseases. Relicts of these developments can still be seen in many eastern old-growth forests, which still contain a generation of large old oaks that grew up in a landscape kept open by native Americans. In fact, native Americans had maintained their influence on the forests for tens of thousands of years, long enough for people and forests to evolve together (Pyne 1982). At the present time, remaining old-growth stands are changing in response to the direct effects of visitors and the indirect effects of management and environmental modification in the surrounding landscape (e.g. Cook 1996).

So what about the perpetual debate about people as part of, or separate from, nature? Semantically, we make a distinction: in ecology and conservation, 'natural' is a separate word that must exclude people to retain its meaning; scientifically, it is useful to think in terms of natural systems as controls or reference points without people, which enable the influence of people to be measured. Historically, however, we know that people have long interacted with the environment and other species. Practically, we know that this will continue. And, ethically, we believe this continued interaction can be desirable if the balance does not shift too far towards people.

# 5

# Biological Interactions and Disturbance: Plants and Animals

## INTRODUCTION

Anthropogenic disturbance of forest environments in recent decades is an increasing cause for concern with regard to conservation of global biodiversity. Indeed, the rate of forest clearance is greatest in areas where the highest endemism of birds (Balmford & Long 1994) and plants (Mittermeier *et al.* 1998) is recorded. The distribution of biodiversity is generally well known but our knowledge of the ecological roles of forest species remains poor, as emphasis has been placed on documenting biodiversity solely by the number and abundance of species and individuals. The 'biodiversity' usually monitored by research ignores biotic interactions and differs from conceptual definitions of biodiversity that include variety at several hierarchical levels (genes, species, habitats, landscapes).

Additionally, species are not equal in the way they contribute to ecosystem functions (such as turnover and productivity) or processes (e.g. pollination, decomposition and herbivory), and there is some degree of ecological redundancy where the functions previously performed by extinct species can be taken over by one or more extant species (Walker 1992; Lawton & Brown 1993). Although vertebrates are dwarfed by invertebrates in terms of species richness and biomass, they remain extremely important in shaping the temporal dynamics of forest structure and composition by mediating ecological processes through a variety of plant-animal interactions. For this reason the impact of anthropogenic disturbances on vertebrate populations in forests, particularly tropical forests, requires careful consideration by forest managers as well as conservation biologists.

Selective logging in the tropics refers to the extraction of commercially valuable trees exceeding a specified minimum size. This accounts for about 3-7% of forest trees, although incidental damage by mechanized logging may result in the destruction of 50% or more trees (Pinard & Putz 1996). In boreal forests where tree species richness is limited to a few species, clearcut harvesting is undertaken and decisions on the size and distribution of clearcut patches have great relevance to the persistence of biotic forest communities (Franklin & Forman 1987). The natural disturbance regime in tropical forests tends to be smaller but more frequent (fallen branches and trees, landslides), although hurricanes can be a major force in some tropical forests, while in temperate and boreal forests natural disturbances are less frequent but more extensive and severe (fire, pest and disease outbreaks, windthrow, etc.). *Consequently, approaches to the ecological management of production forests differ between tropical and temperate regions, with greater emphasis placed on local management in tropical forests, 'while landscape-scale approaches are adopted in temperate regions.*

Disturbances on both local and landscape scales affect vertebrates by altering the abundance and quality of resources such as food or breeding sites, while the composition and stability of whole communities, sometimes encompassing the entire range of species (e.g. the spotted owl in north-western USA and Canada), respond to landscape-scale disturbances such as habitat fragmentation and conversion.

All forest habitats are also subject to natural disturbance regimes that act on spatial and temporal scales, ranging from the local and brief (trampling, grazing, treefalls and gap formation) to regional and prolonged (fires and hurricanes). Understanding the role of vertebrates in shaping the forest community dynamic, and the response of vertebrates to the natural as well as the historic and increasing anthropogenic disturbance regimes at the whole range of spatial and temporal scales, is crucial to the sustainable management of natural forests and the biodiversity they contain.

The purpose of this chapter is to outline the importance of terrestrial vertebrates in plant-animal interactions in tropical and temperate forests, and to describe the impact of disturbances on vertebrates and on the ecological processes they mediate. Although this chapter is limited to a consideration of forest vertebrates, we recognize the important role invertebrates have in mediating forest processes. We concentrate on anthropogenic disturbances that are the cause of much concern with regard to the persistence of forest vertebrates and also focus on tropical forests where vertebrate species diversity is greatest; however, we also draw upon studies from temperate forests. Most studies are conducted at local scales and deal with one or a few species, but in reviewing these studies we attempt to describe emerging regional patterns of responses of vertebrates and vertebrate-plant interactions to local- and landscape-level disturbances.

## DIVERSITY AND DISTRIBUTION OF FOREST VERTEBRATE RESOURCES

### Habitat structure and some generalities of vertebrate responses to resource variability

Vegetational diversity and structural complexity have long been correlated with animal diversity (MacArthur & MacArthur 1961; Karr & Roth 1971; Urban & Smith 1989; Hansen & Hounihan 1996). Vertical habitat complexity and heterogeneity of regeneration phases in old-growth forests, compared with structurally simpler early successional forests, provide a variety of foraging, nesting and roosting sites and diverse microclimatic conditions that support high vertebrate diversity by niche segregation (MacArthur & MacArthur 1961). In northern temperate old-growth forests, elevated light levels in canopy gaps provide vertebrates with highly productive forage patches compared with younger forests that have continuous dense canopies (Alaback 1982). Large standing and fallen dead trees are far more common in old-growth than young regenerating forest (Spies *et al.* 1988) and also provide a range of foraging and nesting opportunities. Foliage volume may be related to avian diversity by affecting the abundance of food resources or nest sites (Mills *et al.* 1991; see also Willson 1993). However, correlation of forest structural complexity with bird species diversity has proved to be highly variable among regions (Hansen & Hounihan 1996) and causal mechanisms linking forest structural complexity with bird diversity remain to be found.

Mobility allows vertebrates to exploit temporally variable food resources and many, particularly mammals, are relatively generalist in their food requirements (Karr *et al.* 1990; Willson 1993). Consequently, vertebrates are able to respond behaviourally to changes in the distribution of ephemeral resources by localized habitat shifts (Levey 1988; Loiselle & Blake 1991), expansion of the home range (Fleming 1992) or switching to alternative food sources (Worthington 1982; Willson 1993). Behavioural responses to regional changes in resource availability can affect local abundance of migratory birds (Stiles & Clarke 1989) and mammals (Ashton 1988; Tutin *et al.* 1997). High mobility buffers forest vertebrates (with some exceptions) from resource-associated breeding failure and extreme fluctuations of population size. However, extensive and prolonged anthropogenic or natural disturbances can affect vertebrate population dynamics by reducing clutch size or causing temporary suspension of breeding (Wiens 1989; Holmes 1990). Some resources have importance disproportionate to their abundance, often because they are available at times of food scarcity, and these may be particularly relevant to breeding success and population viability. Fig trees provide such 'keystone' resources to frugivorous birds and mammals in some tropical forests, and loss of fig trees is predicted to have wide-ranging impacts on the vertebrate communities of these forests (Ter-borgh 1986; Lambert & Marshall 1991).

## Issues of scale

Vertebrates are highly mobile and use resources on much larger spatial scales than plants or invertebrates. Foraging ranges of vertebrates may encompass several hundred hectares or more (Table 5.1), and habitat requirements for nesting and feeding may be in quite different locations. Local disturbances may therefore be less important than large-scale disturbance phenomena in affecting vertebrate communities, and habitat patterns across landscapes should be more relevant to management strategies directed to forest vertebrates. Thus vertebrates are more likely to respond to habitat fragmentation and the heterogeneity and relative abundance of different habitat types than to forest structure or local complexity.

Many vertebrates use resources from a variety of habitat types that occur in natural forested landscapes. These habitat types include a range of successional stages that are topographically, edaphically and hydrologically distinct vegetation zones. Human land uses tend to simplify or homogenize landscapes (Krummel *et al.* 1987). Forest fragmentation leads to habitat isolation, reduced patch size and increased importance of edge effects as well as, more obviously, habitat loss. Single forest patches in anthropogenically fragmented landscapes are unlikely to contain the variety of habitat types and resources previously represented in the continuous forest landscape. Vertebrates that use resources from several forest habitat types are therefore susceptible to changes in resource availability at the landscape scale. Single species operate at a variety of spatial scales during their life history. Thus many small birds defend nesting territories measured in square metres, use foraging ranges extending to several hectares, yet cover many kilometres during seasonal migrations. Even non-migratory species use resources, with varying strengths of interaction, at a range of spatial scales, and incorporating habitat features at multiple scales has been an important planning requirement for conservation of vertebrates such as the spotted owl *Strix occi-dentalis* in the forests of the Pacific North-West of the USA (King *et al.* 1997).

In fragmented landscapes the extent of disruption depends on the scale of fragmentation relative to the mobility of the organisms and on the nature of the habitat matrix in which the fragments are located. Vertebrates that feed on unpredictable patchy and ephemeral resources, such as nectarivores and frugivores, operate on large spatial scales and tend to be more mobile than folivores or insectivores of similar size (e.g. Oates 1987; Fleming 1992). Mobility facilitates movement between patches and consequently offers a degree of protection from the effects of habitat fragmentation, provided the matrix in which resource patches are located is favourable for movement and relatively risk-free. The nature of the habitat matrix and persistence of the resource is particularly important for vertebrates that use temporally and spatially predictable but ephemeral resources that require seasonal migrations between habitat zones (e.g. wild pigs, Ashton 1988; primates, Tutin *et al.* 1997). Seasonal habitat shifts are widespread among birds and mammals in response to climate, the availability of food resources, and nesting requirements (reviewed in Fleming 1992). For example, bears in North America feed on berries in forest habitats but move to streams to feed on salmon during the annual spawning runs (Willson 1993). Sparse berry crops can increase the movement of bears to streams, while a good fruit crop reduces the size of their home range.

Forest patches surrounded by a hostile habitat matrix (or internal fragmentation features such as fences and roads) can impede vertebrate movement that, in addition to limiting access to resources, isolates subpopulations and prevents genetic mixing. By isolating subpopulations fragmentation increases local extinction probabilities by reducing the size of effective breeding population below thresholds of genetic and demographic viability (e.g. Kinnaird & O'Brien 1991). Provision of habitat corridors and careful planning of the land-use regime can increase landscape connectivity and facilitate greater movement of vertebrates between patches. Recent years have seen an increase in studies describing the responses of vertebrates to habitat fragmentation. These have been accompanied by theoretical developments, most notably describing thresholds of fragmentation beyond which the effects of isolation and small patch size increase rapidly (Andrén 1994; With *et al.* 1997).

The concepts of scale and heterogeneity in natural ecosystems are highly relevant to the integration of ecological principles with land management. The size and distribution of habitat patches need to be considered relative to the management objectives, be they conservation of particular species, continued sustainability of ecological processes, or both. Land managers will be aided by remote sensing techniques and geographic information systems (GIS) in the assessment of the suitability of landscapes to support high biodiversity (Short *et al.* 1996), although these approaches need to be combined with studies of life history traits (Hansen *et al.* 1993) and species requirements. Variation in the frequency and size of natural disturbance events should be accounted for in management planning by incorporating or mimicking natural disturbance regimes at several spatial and temporal scales. Such planning needs to take a hierarchical perspective to match the mobility and requirements of biological entities that vary in size from small arthropods to herds of large mammals.

## VERTEBRATE RICHNESS AND DIVERSITY IN PRIMARY AND DISTURBED FORESTS

Many studies have examined how habitat transformation affects the richness and abundance of forest animal communities (Thiollay 1992; Andrade & Rubio 1994; Laurance 1994; Malcolm 1994). Results from these studies are highly variable, reflecting the complex interplay of abiotic and biotic factors acting at different spatial and temporal scales on community structure and composition. Disturbance itself is qualitatively and quantitatively highly variable; deforestation, the most widespread anthropogenic disturbance, causes habitat fragmentation (with associated impacts of edge effects and isolation), microclimatic changes and alteration of habitat structure. These changes may further lead to increased susceptibility to fire (Holdsworth & Uhl 1997; Uhl 1998), windthrow (Laurance 1997), pests and pathogens, or invasive exotic species. This interplay of biotic and abiotic complexity, coupled with a scarcity of long-term datasets, makes it difficult to distinguish patterns describing the effects of habitat alteration on biodiversity and ecological processes (Kruess & Tscharntke 1994; Pimm & Sugden 1994; Tilman & Downing 1995).

Vertebrates themselves may be instrumental in structuring the physical nature of the habitat, which in turn has important positive and negative effects for other species occurring within the environment. Particular vertebrate species may have disproportionate roles as agents of habitat structural manipulation. Jones *et al.* (1994, 1997) term such organisms 'physical ecosystem engineers'. Ecosystem engineering plays a major role in the structure and function of ecosystems by controlling the availability of resources through the physical modification, maintenance or creation of habitats (Jones *et al.* 1997). In temperate forests, beavers have an important function as ecosystem engineers through the damming of forest streams, which increases habitat and resource heterogeneity (Naiman *et al.* 1988). Squirrels can also be important through the damage they cause to trees. In tropical systems, elephants and herds of ungulates change habitat physical structure by damaging trees and trampling the undergrowth (e.g. Lewis 1991). The effects of some engineers may be dramatic in proportion to their size or abundance. Woodpeckers excavate nesting cavities in dead wood that are later used by a wide variety of birds, mammals and invertebrates unable to excavate cavities themselves (Harris & Silva-Lopez 1992). Recognizing the role of these species as ecosystem engineers in forest systems is crucial for ecologically sustainable forest management.

Particular organisms can also have key roles in the food web. The role of predation in structuring vertebrate communities has largely been unexplored, though recent work indicates that predation may indeed be important (Terborgh 1988; Brosset 1990; Fonseca & Robinson 1990; Karr *et al.* 1992). As top predators, large carnivores are susceptible to habitat fragmentation or conversion (as well as hunting) owing to their small population sizes, large foraging range requirements and relative inflexibility in their food preferences (Terborgh & Winter 1980; Fonseca & Robinson 1990). The effects of the loss of top predators are not clear, although island communities or recently fragmented habitats that cannot, or no longer, support populations of large predators indicate that densities of herbivores and seed-predators increase, with further effects cascading through the community. The El Verde forest of Puerto Rico, for example, has unusually high abundances of lizards and amphibians. High densities of small mammals on Barro Colorado Island in Panama are thought to be due to the absence of regulation by large predators (Terborgh & Winter 1980). Changes in trophic structure can shape plant communities by altering patterns of herbivory and seed predation. Thus on Barro Colorado Island, recruitment of *Dipteryx panamensis*, a large canopy tree, is much reduced compared with the mainland due to unusually high seed predation pressure by small mammals (De Steven & Putz 1984).

The existence of forests themselves may be a temporal dynamic regulated by the interaction of abiotic and biotic disturbances. These disturbances can, in combination, cause major environmental changes that each on its own cannot achieve. The interaction of disease, fire and grazing by large mammals appears to be responsible for temporal changes in savannah woodland cover in Africa (Pellew 1983; Dublin *et al.* 1990). For example, woodland growth in the Mara region of the Serengeti is facilitated by wildebeest, which maintain fires at low frequency and intensity by grazing the grass fuels. Increased fire following reduction of wildebeest populations by disease can lead to a contraction of forested land, although this is exacerbated by population increases of elephants (due to a policy of elephant conservation) that inhibit forest regeneration by browsing of tree seedlings (Dublin *et al.* 1990).

The rest of this section reviews vertebrate diversity and abundance in anthropogenically disturbed forest. The response of bird and, to a lesser extent, mammal communities to disturbance has been well studied throughout temperate and tropical regions, permitting some generalizations to be drawn. In contrast, very few studies have investigated impacts on amphibians and we could find only one referring specifically to reptiles.

## Birds

Forest degradation and fragmentation has been repeatedly shown to reduce bird diversity and richness, increase the dominance of the commonest species, and favour gap and forest edge specialists at the expense of forest interior species, which often decline in density or become locally extinct (Johns 1991; Thiollay 1992; Bierregaard & Stouffer 1997; Canaday 1997; Marsden 1998). In-sectivores (Johns 1991; Thiollay 1992; Bierregaard & Stouffer 1997; Canaday 1997; Marsden 1998) and to a lesser extent large frugivores (Thiollay 1992; Bierregaard & Stouffer 1997) appear to be most sensitive to disturbance. Among the insectivorous birds, leaf litter insectivores are often the most severely affected (Bierregaard & Stouffer 1997; Canaday 1997), probably reflecting changes in invertebrate abundance and distribution following drying of the leaf litter in fragmented forests (Wolda 1978; Kapos 1989; Canaday 1997). Nectarivores appear to be more resilient to forest degradation and fragmentation, and little change in behaviour or abundance of hummingbirds has been noted in

Central or South America. Some large tropical birds, such as hornbills and parrots, are thought to be tolerant of recent logging (Lambert & Marshall 1991; Marsden 1998), but these species are nomadic, long-lived and require large trees for nesting, and their long-term viability in logged forests has yet to be assessed (Marsden 1998).

In Amazonian forest fragments common frugivores were heavily affected while uncommon species showed no apparent response (though small sample sizes of rare species limit interpretation) (Bierregaard & Stouffer 1997). Such results offer some promise for conservation of avian biodiversity, but there may be serious implications for seed dispersal and plant regeneration (see also Silva *et al.* 1996). Recolonization of Amazonian forest fragments by insectivores appears to be a function of the quality of the surrounding habitat matrix (Bierregaard & Stouffer 1997), which may have great importance in maintaining the long-term biodiversity and resilience of insectivore communities. In contrast, little recolonization of fragments by frugivores occurred regardless of the nature of the surrounding matrix, neither was there any compensation of the seed dispersal function in fragments by frugivores associated with secondary growth. Retention of living and dead residual trees in clearcuts or large forest gaps has been frequently proposed as a strategy that benefits wildlife (Raphael & White 1984; Hunter 1990; Payne & Bryant 1994). Retention of standing trees can increase the use of clearcuts by birds (Dickson *et al.* 1983; Marcot 1983; Niemi & Hanowski 1984) and increase local avian diversity (Dickson *et al.* 1992). However, bird species differ in their habitat requirements and there is concern that management for high diversity may be detrimental to habitat specialists or rare species (Harris 1984; Yahner 1988). Merrill *et al.* (1998) addressed this concern and found that residual trees in Minnesota aspen clearcuts do benefit several avian species of high conservation priority but cautioned that the value of residual stands is relatively minor compared with large intact forests for supporting endangered forest bird species.

Fragmentation of the northern boreal forests of Canada, which has one of the richest avian breeding sites in North America

(Robbins *et al.* 1989), can reduce the quantity and quality of the available breeding habitat (Saunders *et al.* 1991) with consequent effects on the avian community. Predation and brood parasitism of forest birds have been shown to increase in anthropogenically disturbed or fragmented forests (Andrén *et al.* 1985; Wilcove 1985; Andrén & Angelstam 1988; Yahner 1988, 1996; Burkey 1993; Andrén 1995; Robinson *et al.* 1995; Bayne & Hobson 1997) and along forest edges (Andrén *et al.* 1985; Santos & Telleria 1992; Paton 1994; Andrén 1995). In a quantitative review of these studies Hartley and Hunter (1998) concluded that predation rates are much higher in deforested areas but that the strength of edge effects is a function of landscape composition, notably the extent of remaining forest cover and the quality of the non-forested matrix. Nest parasitism by brown-headed cowbirds, *Molothrus ater*, which proliferated following human settlement and forest fragmentation, has severely affected the fauna of North American forests (Brittingham & Temple 1983; Robinson & Wilcove 1994). However, birds in boreal forests are more likely to be affected by fragmentation caused by agricultural development than fragmentation due to logging (Hartley & Hunter 1998), as the latter is more likely to mimic natural disturbance regimes caused by fire or insect tree-pest outbreaks (Hansson & Angelstam 1991; Hansson 1992; Bayne & Hobson 1997; see also Santos & Telleria 1992). Finally, fragmentation or decline of temperate forests can affect the populations of tropical migrants (Robbins *et al.* 1989; Robinson & Wilcove 1994; Robinson *et al.* 1995) and coordinated forest management across widely separated regions is needed to prevent the decline of migratory species.

## Mammals

At intensities of 70-90 $m^3ha^{-1}$ (and sometimes reaching 120$m^3ha^{-1}$), selective logging in South-East Asian dipterocarp forests is among the highest in the tropics. Despite this, most mammals studied are able to persist in logged forest (civets, Heydon & Bulloh 1996; primates, Johns 1992; ungulates, Heydon 1994—quoted in Heydon & Bulloh 1997), although carnivores may be more adversely affected than herbivores and omnivores. Diet flexibility may contribute to the persistence of mammal populations in logged forest, and the ability of frugivorous primates to include leaves in the diet appears to be correlated with their survival (Johns & Skorupa 1987). High-energy requirements, associated with small size, may restrict mammalian diets to foods that are easily digestible and have high calorific content. Thus small mammals (e.g. mousedeer, Heydon & Bulloh 1997) are thought to be more susceptible to changes in resource quality or abundance.

There are contrasting responses to disturbance between insectivorous and frugivorous bats, presumably a reflection of their feeding guilds. Insectivorous bats are not affected by losses of canopy cover as abundance of their aerial insect prey does not vary across disturbed and undisturbed forest habitats or even in surrounding pastureland (Fenton *et al.* 1998). Flying foxes feed on forest fruits and are more susceptible to forest degradation (Cox *et al.*1991), presumably due to the loss of fruiting trees in logged forest (Johns 1992).

The resilience (i.e. rate of population recovery) of mammal communities in logged forests appears to be high. Studies in peninsular Malaysia suggest that the initial effects of extraction of more than 50% of trees on primates and other mammals may be considerable but that most species present prior to disturbance had persisted or recolonized 12 years later (Johns 1992).

However, as with birds, the nature of the surrounding habitat matrix has a large influence in determining the composition of forest fragments and Malcolm (1997) showed that primary forest fragments in Amazonia surrounded by secondary forest maintained much higher densities and diversities of small mammals than fragments surrounded by pastureland. Some studies have reported marsupials (Fonseca & Robinson 1990) and rodents (Malcolm 1997) to be even more common in secondary than primary forest.

The structural and microclimatic conditions of secondary forest may resemble closely that of natural treefall gaps where rodents preferentially forage (Forget 1997).

Hunting pressure can be high in logged forests, the preferred species being large birds and mammals (Redford 1992; Bennett & Dahaban 1995; Rumiz *et al.* in press), although rodents are also targeted (Forget & Milleron 1991). These are often the animals that have important roles as seed predators and dispersers (see below) and there is some evidence that tree recruitment may be negatively affected by changes in the composition of mammal communities through hunting pressures (see Guariguata & Pinard 1998).

## Amphibians

Strong site fidelity and limited dispersal capabilities appear to make amphibians particularly sensitive to the effects of logging and habitat fragmentation (Bradford *et al.* 1993; Blaustein *et al.* 1994; Marsh & Pearman 1997), although there is evidence suggesting that observed declines in amphibian richness immediately after a disturbance event is reversed in following years and richness may ultimately even exceed predistur-bance levels (Tocher *et al.* 1997).

Changes in forest structure can lead to the drying of the habitat and increased daily temperature variation (Bastable *et al.* 1993). Most terrestrial amphibians are highly sensitive to these changes due to physiological and developmental dependence on high atmospheric humidity and abundant water. However, some species (e.g. *Eleutherodactylus* spp.) deposit their eggs on land and undergo direct development to the adult form within the egg. Such life-history features allow these species to achieve very high densities and become distributed widely throughout the forests in which they occur (Reagan *et al.* 1996).

More generally, amphibian species richness has been correlated with habitat characteristics such as humidity, leaf litter thickness (Fauth *et al.* 1989; Woinarski & Gambold 1992) and understorey density (Pough *et al.* 1987; Dupuis *et al.* 1995), and amphibian abundance positively correlated with forest fragment size and proximity to larger forest tracts (Marsh & Pearman 1997). The availability of a deep litter layer that maintains high moisture levels appears to be particularly important for terrestrial nesting amphibians (Hodl 1990). Thus shady dense second growth appears to be more accessible to amphibians than exposed pasture and, consequently, the nature of the surrounding matrix affects rates of amphibian colonization of disturbed areas. Like other animal groups, amphibians exhibit highly species-specific responses to disturbance (Tocher *et al.* 1997).

## Reptiles

In the almost total absence of published studies we make some tentative predictions about the response of reptiles to anthropogenic disturbances based on physiological and ecological attributes. Reptiles are poikilothermic and therefore depend on external sources of heat to raise body temperature and, consequently, they have low metabolic rates and energy requirements.

Canopy opening by disturbance changes the understorey microclimate by raising daytime temperatures in canopy gaps. This is likely to benefit heliothermic reptiles by extending their activity periods (Vitt *et al.* 1998). As with insectivorous birds, drying of the leaf litter may affect smaller lizards due to reduced arthropod availability.

However, compared with homeothermic birds and mammals, reptiles are not likely to be limited by food or water availability as many are able to survive long periods between feeding events. Increased abundance of rodents in gaps (Forget 1997) and secondary forest (Malcolm 1997) may even increase food availability for large reptiles such as snakes and teiid lizards (Vitt *et al.* 1998).

This could increase the abundance of predaceous reptiles by immigration and population growth, which may have a cascade effect on other forest species (Vitt *et al.* 1998).

Furthermore, fragmentation can concentrate forest birds into smaller areas, thereby raising the density of prey such as eggs and chicks. Smaller patch sizes and increased edge effects may facilitate the movement of reptiles normally associated with open pasture into forested areas. Fragmentation could affect populations of forest turtles if forest patches become too small to include streams or areas of seasonally available standing water.

Thus we predict that reptiles are less likely than birds or mammals to be affected by conversion of primary forest to secondary forest or to a series of forest fragments. Food and water (except perhaps for turtles) are unlikely to be limiting as their metabolic requirements are relatively low. Predation by birds and mammals may be more relevant in structuring reptile communities, but little is known about the changes in the foraging preferences of predators in disturbed habitats.

## VERTEBRATE-MEDIATED ECOLOGICAL PROCESSES AND IMPACTS OF DISTURBANCE

Ecological interactions have critical importance in the maintenance of biodiversity and ecological functions such as productivity and turnover. Mutualistic interactions are extremely common and include interactions between all major organismal groups, they have been largely neglected in the ecological literature, with greater emphasis placed on competition and predation. Many interactions may be quite unspecific with a great degree of ecological redundancy, such that the loss of a single species member of a mutualism results in little change in the overall outcome of the interaction. The extent of redundancy and the resistance of ecological interactions to anthropogenic disturbances are poorly known, yet of critical importance to forest managers in determining extractive mechanisms and limits. Vertebrate-plant interactions should be of interest to both forest managers and conservation biologists owing to the roles of vertebrates in plant recruitment and in regulating forest structure and composition through pollination, seed dispersal, and seed predation and herbivory. This section reviews these four plant-vertebrate interactions in forest habitats and considers disruption of these interactions by anthropogenic disturbance.

### Pollination

Pollination by vertebrates is almost non-existent in northern temperate forests where most species are wind-pollinated. Indeed there is a trend of increasing wind pollination with increasing latitude (Regal 1982). In tropical forests, high tree diversity and wide spacing between conspecifics limit the efficiency of wind pollination and tree species in lowland tropical forests are almost all pollinated by animals (including insects). To attract vertebrate pollinators plants need to produce nectar in quantities much greater than that required for insect pollination. The energy costs of nectar production to plants can therefore be considerable. Thus nectar produced by vertebrate-pollinated plants tends to be relatively dilute. Vertebrate-pollinated plants are therefore most abundant in wet tropical climates where water does not limit nectar production, and are rarely found in dry forest or savannah habitats. Among vertebrates, bats, birds and a few non-flying mammals are the only known pollinators (Bawa 1990); although they pollinate only a minority of plants (6% of canopy tree species in tropical moist forest at La Selva, Costa Rica; Bawa 1990), they remain crucial for the successful reproduction of many large forest trees and a much greater variety of forest understorey plants (Hammond *et al.* 1996; Corlett 1998; Brewer & Rejmanek 1999).

Well-documented cases of non-flying mammal pollination involve 59 mammal species distributed among 19 families and six orders (Carthew & Goldingay 1997). The marsupials of Australia (12 species) and primates of Madagascar (9 species) and South America (10 species) are the best-known examples. However, the importance to plants of non-flying mammals as pollinators is debatable, as many of these plants are often served by other vertebrate and insect pollinators. Only in Madagascar do non-flying mammals (lemurs) act as sole pollinators of the plants they visit (Nilsson 1993; Kress *et al.* 1994), and the precipitous decline of some lemur species may cause an associated decline of the plants they pollinate (e.g. the traveller's tree *Ravenala madagascariensis* pollinated by the endangered black-and-white ruffed lemur). Selective exclusion experiments have demonstrated that marsupials are at least as effective as birds in pollinating *Banksia* in Australia (Goldingay *et al.* 1991; Carthew 1993) and *Protea* spp. in South Africa (Wiens *et al.* 1983); indeed, declining small-mammal activity at *Banksia* inflorescences due to habitat fragmentation resulted in a sharp decrease in seed set of *Banksia goodii* (Lamont *et al.* 1993). However, documented cases of reduced seed set due to declines in mammal pollinators are rare. In most cases, evidence for effective pollination by non-flying mammals is largely circumstantial and, while non-flying mammals may feed at flowers and occasionally pollinate them, this service must be balanced with the damage they cause (Janson *et al.* 1981).

Bats pollinate species belonging to several tropical plant families including Bombacaceae *(Pseudobombax)*, Mimosaceae *(Parkia)*, Passi-floraceae *(Passiflora)* and Caesalpiniaceae *(Bauhinia)*; on oceanic islands, such as Samoa, the majority of the dominant canopy tree species may be pollinated by bats (Cox *et al.* 1991). A number of economically important tropical trees rely on bats for pollination, including durian, banana, neem and timber species of *Eucalyptus* and several species of palm (Fujita 1991—quoted in Allen-Wardell *et al.* 1998). Although it is energetically expensive to produce bat-attracting flowers rich in nectar, plants benefit from extensive pollen transfer as bats forage over large areas (Heithaus *et al.* 1975). Habitat change, hunting and the introduction of exotic species is causing the decline of bats across the tropics and several species are threatened with extinction. This is of particular concern on oceanic islands, where alternative pollinators for endemic tree species are rare or absent (Cox *et al.* 1991).

In the Neotropics hummingbird pollination is widespread among several plant families (Regal 1982), while old-world pollinators include a much greater variety of birds, primarily sunbirds but also leafbirds, lorikeets, white-eyes and honey-eaters. However, birds are rarely the exclusive pollinators of plants (but see Ford & Paton 1985) and their importance relative to insect pollinators is not well resolved. Declines in fruit set due to temporary absence of hummingbirds have been noted, even where other invertebrate pollinators were present (Waser 1979), but obligate pollinator dependency by plants on hummingbirds or sun-birds has not been documented. Rather, these birds may be important as biological 'insurance' where temporal fluctuation of invertebrate pollinator populations may be severe (Aizen & Feinsinger 1994).

Loss of vertebrate pollinators resulting from forest conversion or fragmentation is only likely to affect plant seed production if the plant-pollinator interaction is highly specialized. Most plants are pollinated by a suite of mainly general-ist pollinators (Waser *et al.* 1996) and are unlikely to suffer a large decline with the loss of one or a few closely related pollinator species. Nevertheless, in tropical floras specialist interactions among animals and plants do exist. On the small isolated Caribbean island of Tobago the absence of hermit hummingbirds resulted in reduced seed production of *Mandevilla hirsuta* (Apocynaceae) compared with the neighbouring island of Trinidad where the birds were present (Linhart & Feinsinger 1980). In contrast, there was no difference across islands in seed production of *Justica secunda* (Acanthaceae) that is pollinated by a more general array of hummingbirds. In fact most hummingbirds appear to be quite plastic in their habitat requirements and regularly use edges and move between primary and secondary habitats to take advantage of a wide variety of floral resources (Feinsinger 1976; Bierregaard & Stouffer 1997); there is also evidence that sunbirds, which in the Palaeotropics fill a similar ecological niche to hummingbirds, are more abundant at flowering trees in disturbed secondary forests (Marsden 1998).

However, predicting the impacts of habitat and landscape modification on plant-pollinator interactions is difficult as both the behaviour and abundance of pollinators may change as a result of changing resource distribution (Ghazoul *et al.* 1998). Pollinators may also depend on resources that are far from the plants they pollinate. In peninsular Malaysia, for example, the fruit bat *Eonycteris spelea* is the exclusive pollinator of durian trees but feeds primarily in coastal mangroves. The bats visit, and pollinate, durian *en route* to the mangroves from their limestone-cave roosts (Start & Marshall 1976). Destruction of either the caves or the mangroves could affect the bat population and hence the pollination of durian.

## Seed predation

Predation of seeds by animals has been proposed as a mechanism for the maintenance of the high tree species diversity in tropical forests (Janzen 1970; Connell 1971). This theory suggests that density-dependent consumption of seeds by mammalian seed predators prevents domination of the vegetation by species that have competitively superior seeds or greatest seed production. Terborgh (1988) suggested that vertebrate seed predators selectively prey upon the largest tree seeds, and where carnivores are eliminated following perturbations forest composition might shift to smaller-seeded species. There is some evidence to support this prediction from forests in Panama and Mexico (De Steven & Putz 1984; Dirzo & Miranda 1990). In Panama, high densities of small mammals on Barro Colorado Island due to an absence of large predators (Terborgh & Winter 1980) is thought to be the cause of recruitment failure among the large canopy tree *Dipteryx panamensis* due to the seed predation pressure exerted by the small mammals (De Steven & Putz 1984).

Some studies show seed predation by vertebrates (most often rodents) to be higher in mature forest compared with early successional forest (Janzen 1985a; Santos & Tellería 1992; Aide & Cavelier 1994; Osunkoya 1994), while other studies found the opposite (Uhl 1987; Hammond 1995). Rodents tend to avoid large areas of open forest or pasture probably due to the lack of protective cover and the scarcity of food resources (Price & Jenkins 1986). However, seed predation intensity under different disturbance regimes varies greatly with the species of seed plant (Willson & Whelan 1990; Terborgh *et al.* 1993; Osunkoya 1994; Chapman & Chapman 1996; Holl & Lulow 1997) making it difficult to draw generalizations to guide forest restoration programmes.

However, where seed predator densities are high, commercially available low-value grains might be used to satiate the seed predators in order to minimize mortality of the desired species. Use of these methods should be based on a sound knowledge of the factors that limit the population dynamics of the predator species, otherwise the availability of superabundant food resources leads to rapid population growth and subsequent heavy seed mortality in later years.

## Seed dispersal

Seed dispersal has been favoured by evolution because it decreases the probability of seed predation, which is often disproportionately high close to parent trees, and increases the probability that seeds find a suitable establishment site (Howe & Smallwood 1982).

Thus seeds have become adapted for dispersal by a variety of vectors, the most common of which are wind and animals, and the role of animals in the dispersal of seed and fruit has received much attention in the biological literature (see review by Sallabanks & Courtney 1992). Seeds carried by animals often have edible appendages (e.g. *Acacia* spp., *Anacardium* spp.) or are enclosed within fleshy fruits that are consumed and the seeds ejected later. In some cases the seeds themselves are harvested but dropped, or stored and later forgotten or abandoned.

In Neotropical forests, 50-90% of the canopy trees and nearly all shrubs and subcanopy trees bear fruit adapted for animal dispersal, while a smaller but still substantial proportion of species in Palaeotropical forests are also dispersed by animals (35-48% and 70-80%, respectively) (Howe & Smallwood 1982; Howe 1986). Birds, bats and scatter-hoarding rodents are the major dispersers of tropical tree seeds (Hammond *et al.* 1996; Corlett 1998; Brewer & Rejmanek 1999), while other seed dispersers include primates, fish in Amazonian forest, tortoises, some herbivorous lizards and even a frog (Stiles 1989, 1992).

In temperate coniferous and deciduous forests, seeds may be more often dispersed by wind or by cache-hoarding mammals and birds (Fenner 1985). However, this dichotomy between temperate and tropical forest dispersal mechanisms has been challenged by Wheelwright (1988), who asserts that interactions between plants and their fruit-eating bird dispersers are as complex and ecologically important in temperate regions as they are in the tropics. Howe and Smallwood (1982) state that more than 60% of temperate trees have animal-dispersed seeds and record a significant negative correlation between annual precipitation and the percentage of wind dispersal. Nevertheless, there are more species of fruit-eating birds in the tropics and they comprise a larger fraction of the avian biomass (Terborgh 1986). Only in the tropics are there bird species that rely exclusively on a diet of fruit (e.g. fruit pigeons *Ptilinopus* spp. in Australia and *Treron* spp. In South-East Asia), and most birds with a predominantly fruit diet are tropical species.

Seed dispersal syndromes are thought to reflect coevolutionary trends between plants and vertebrate (bird and mammal) seed dispersers. Bird-dispersed fruit are small and red, black, blue or purple (van der Pijl 1969), are conspicuous by sight rather than smell, and persist on the tree until removed by a disperser (Janson 1983; Howe 1986). Mammal-dispersed fleshy fruits are large, green or brown (van der Pijl 1969; Willson *et al.* 1989), emit odours, and frequently abscise shortly after ripening (Janson 1983; Howe 1986). However, the existence of clearly defined seed dispersal syndromes has recently been questioned (Tamboia *et al.* 1996) and consumption of a particular type of fruit by an animal does not imply that the fruit is adapted for dispersal by that animal (Janzen 1985b). The emerging view is that most interactions between plants and avian seed dispersers, in both tropical and temperate regions, are generalist, opportunistic and largely inefficient (from the plant's perspective), and there are few examples of tightly coupled coevolved mutu-alistic interactions. One well-documented exception is the commercially important brazilnut *Bertholletia excelsa*, which is entirely dependent on agoutis to gnaw open the seed capsules and release the seeds (Peres & Baider 1997).

Animal behaviour can be crucial to the success of plant dispersal and recruitment. Animal seed dispersers often place seeds in sites suitable for germination, and seeds of some species (e.g. oak) will not germinate unless they are buried (Griffin 1971). Burial further reduces the probability that the seeds will be found by seed predators (Vander Wall 1993). Bats may reduce the suitability of fruit to insect seed predator attack, leading to increased survival of seeds dropped at a bat roost (Janzen 1982).

Animal digging, burrowing and tunnelling (by armadillos, coatis, porcupines, pigs, gophers) can bury seeds or unearth buried seeds. At Mount St Helens, for example, the activity of surviving gophers in bringing soil and propagules to the surface of the sterile ash deposits facilitated plant establishment and the start of forest regeneration (Andersen & Mac-Mahon 1985). Secondary dispersal of wind-dispersed seeds of *Pinus jeffieyi, Pinus ponderosa* and *Pinus contorta* by chipmunks and other rodents in the Sierra Nevada is instrumental in moving seeds up to 69 m away from the parent tree compared with only 12 m by wind dispersal alone (Vander Wall 1992). Mockingbirds, doves and land iguanas have been essential in facilitating regeneration of the tree *Bursera graveolans* (Burseraceae) on the Galapagos Islands after the elimination of introduced goats. The *Bursera* seeds fail to germinate directly beneath adult trees, and recovery of the population to levels prior to the introduction of goats is entirely dependent on the native seed dispersers (Clarke & Clarke 1981). Vertebrate dispersal agents may further benefit plants by dispersing mycorrhizal fungal spores in addition to seeds, as in the Pacific North-west where squirrels disperse the obligate symbiotic mycorrhizal fungi of conifer seedlings (Maser & Maser 1988).

Dependence on animals for seed dispersal exposes the plants to the risk of dispersal failure should seed vectors become rare or extinct. The population structure of the tambalocoque tree *Sideroxylon sessiliflorum* (Sapotaceaea) on Mauritius is thought to have been affected by the extinction of the dodo (Temple 1977, 1979; but see Witmer & Cheke 1991), which dispersed the seeds and whose elimination doomed the tree to extinction as one of the 'living dead' (Janzen 1988). The likely cause of the dramatic decline of colonization rates of many fleshy fruited species in the remnant rain forests of the Mascarene Archipelago, in which Mauritius is located, is thought to be due to the extinction of over 50% of its land vertebrates in recent decades (Cheke 1987; Thebaud & Strasberg 1997). Cox *et al.* (1991) describe preliminary observations of pollination and seed dispersal failure for several bat-pollinated and -dispersed plants on the island of Guam where bats have suffered severe declines in recent years. Indeed, Fujita and Tuttle (1991) note that at least 289 plant species (according to current published information) depend to varying degrees on large populations of flying foxes for their propagation, and the fact that flying foxes are increasingly threatened is a cause for broad concern.

Selective logging and fragmentation of forest habitats have a large impact on vertebrate abundance and behaviour, which in turn is likely to affect seed dispersal of many animal-dispersed tree species. Forget and Sabatier (1997) reported that large frugivores avoid recently created gaps, and many bird species do not move into central regions of large gaps (Gorchov *et al.* 1993). However the responses of species to fragmentation can be highly variable and the impact on forest tree regeneration is often difficult to predict. Frugivorous birds may decline sharply after fragmentation (Terborgh & Winter 1980) leading to decreased seed dispersal (Santos & Tellería 1994), while small mammal populations remain stable or even increase in abundance (Santos & Tellería 1994; Malcolm 1997). However, the behaviour of small mammals may alter as the landscape is transformed into a primary and secondary forest mosaic. Terrestrial mammals readily move between primary and secondary forests (Malcolm 1997) and may promote regeneration by dispersing primary forest tree seeds into degraded secondary forest (Silva *et al.* 1996; Lamb *et al.* 1997), although seeds can also be transported in the opposite direction and alter the nature of primary forest succession.

Management of animal seed-dispersers may be necessary to accelerate regeneration of degraded areas (Silva *et al.* 1996; Lamb *et al.* 1997) and to promote the establishment of native plant species. Availability of hollows, snags, perches and nest boxes increases the abundance and diversity of birds (McClanahan & Wolfe 1993) and bats (Fenton *et al.* 1998), thereby preserving dispersal functions and accelerating forest regeneration (McClanahan 1986). Where vertebrates choose to sit while processing seeds is critical to the reproductive success of plants and appropriate provision of perches and nest boxes can encourage frugivorous birds to process (and defecate) consumed seeds in degraded sites. However, if tightly coupled mutualisms exist between a plant and animal, then specific management strategies may be required to ensure the continued existence of both.

## Herbivory

Vertebrate herbivores are instrumental in determining the vegetational composition of forest systems. The role of the African elephant in maintaining the balance between savannah and woodland in Africa is well documented (Pellew 1983; Dublin *et al.* 1990; Ruess & Halter 1990) and the impact of grazing and browsing mammals on temperate forests has been reviewed by Gill (1992a,b). The loss of large herbivores from Europe in the Pleistocene is thought to have led to the transformation of short grasslands to less productive forests with concomitant extinctions of many small mammals dependent on the more nutrient-rich vegetation (Owen-Smith 1987, 1989). More recently, the elimination of rabbits by the disease myxomatosis in Britain has resulted in many grassland areas reverting to scrub woodland. Overgrazing by deer in European and North American woodlands causes a decline in native woodland species, facilitates the spread of invasive weeds (Koh *et al.* 1996), prevents regeneration of woody species (Ross *et al.* 1970; Tilghman 1989; Trumball *et al.* 1989; Cooke & Lakhani 1996) and may result in local extirpation of herbaceous plants (Augustine & Frelich 1998). Moderate grazing can lead to plant species replacements. In North America, preferential feeding of white-tailed deer on pine *Pinus strobus* results in its replacement by birch *Betula papyrifera*, which has a greater resistance to browsing (Ross *et al.* 1970). Moderate or low levels of grazing can also result in greater vegetational diversity than either overgrazing or an absence of grazing (Mitchell & Kirby 1990). Thus, in tropical Mexican forests the defaunation of large mammal herbivores such as tapirs, peccari and deer by hunting and illegal trading led to increased density but decreased diversity of seedlings and saplings (Dirzo & Miranda 1991).

Grazing represents a nutritional drain on plants and the ability to survive sustained grazing pressure may depend on soil quality and efficiency of nutrient capture. The response of *Colophospermum mopane* trees to high levels of elephant herbivory in *C mopane*-dominated woodland in Zambia was related to the availability of soil nutrients; poor soil quality combined with elephant herbivory increased plant species diversity and reduced dominance by *C. mopane* (Lewis 1991). Periodic large-scale diebacks were also suggested to be a function of soil nutrient depletion from sustained grazing by elephants in these woodlands.

By browsing and debarking seedlings, saplings and trees, large mammalian herbivores can influence forest plant diversity and abundance, which in turn affect the resource and habitat quality and quantity for a range of other plant and animal species. Woodland degradation in Botswana by African elephants resulted in substantial changes in bird species composition, although bird diversity was unchanged (Herremans 1995), while culling of hippopotamus in Uganda allowed regeneration of plant diversity and consequential increases in buffalo, elephant and waterbuck densities (Hunter 1992). In tropical wet forests, large mammal densities are considerably lower than in African savannah forests and vertebrate herbivores do not have a major impact on habitat structure or composition.

## VERTEBRATE INVADERS AS MEDIATORS OF CHANGE

Vertebrates themselves can be agents of anthropogenic disturbance where exotic species are released into new environments. Although most introductions fail, some succeed; of these, a few can cause large changes in the composition and structure of native forests. A familiar example in the UK is the introduction and subsequent spread of the grey squirrel *Sciurus carolinensis* at the end of the nineteenth century. Aside from eliminating the native red squirrel *Sciurus vulgaris* from much of its former range in the UK, the grey squirrel attacks hardwood trees by stripping bark, which leads to scars in the wood (and even death by ring-barking), and by damaging crowns.

Invasions of alien predators can devastate naïve prey species. The introduced brown tree snake *Boiga irregularis* has been responsible for the extinction or decline of 10 species of native birds on the island of Guam, including the Philippine turtle dove *Streptopelia bitorquata*, an important disperser of seeds into degraded vegetation remnants (Savidge 1987). Diseases spread by the introduction of exotic vertebrates is cited as a cause of the extinction of native Hawaiian birds, although habitat changes and predation by introduced grazing mammals and rats may be more important causes (Pimm 1991). The spread of alien plants may be facilitated by native or alien generalist seed dispersers or pollinators. Thus the spread of the Brazilian pepper *Schinus teribinthi-folius* in south Florida was facilitated by the American robin *Turdus migratorius* and the introduced red-whiskered bulbul *Pycnonotus jocosus* (Ewel 1986). North American mammals readily incorporate the fruit of exotic *Prunus, Malus, Pyrus* and other tree species into their diets (Willson 1993). Likewise, goats and pigs in Hawaii have been responsible for destruction of native vegetation, the creation of open areas and the dispersal of exotic plant seeds through their dung, all factors that facilitate further invasion by exotic plants (Loope & Stone 1996).

The importance of scale and historical association in determining ecological outcomes of species introductions is illustrated by the impact of red squirrels *Tamiasciurus hudsonicus* introduced to Newfoundland on the native subspecies of red crossbill *Loxia curvirostra percna*. The squirrels, introduced as prey for pine martens, quickly reached high densities by outcompeting Newfoundland red crossbills for the seeds of black spruce. The Newfoundland red crossbill is now close to extinction, though the mainland subspecies continues to coexist with the squirrels because of evolutionarily derived differences in habitat and food preferences, a result of a history of coexistence, and because mainland populations are able to travel widely over a largely forested landscape in search of new food resources (Pimm 1990). Ecologically sensitive forest management must therefore consider broad spatial (i.e. landscape) and historical patterns in order to minimize the impacts of introduced predators and competitors and the spread of generalists into forested areas at the expense of obligate forest-interior species.

## ECONOMIC VALUE, HUNTING AND TOURISM

Vertebrate conservation by landscape management and protected area design has been discussed in this chapter and is elaborated further in Volume 2, Chapter 4. Successful conservation depends on integration of human needs and perceptions with those of conservation at local and regional scales. In tropical and temperate regions vertebrates represent a valuable source of income and are widely exploited as such. Exploitation may be destructive, such as hunting and collection for trade, or unobtrusive when income is generated merely through aesthetic or existence value. Vertebrate populations are susceptible to rapid decline following destructive exploitation as initial population sizes and population growth rates are often low. While attentive management of exploited populations is needed, incentives for sustainable use of vertebrates are also required. Here we briefly highlight two approaches particularly relevant to vertebrates where economic incentives are used to promote conservation.

As logging enterprises make large areas of previously remote forest accessible, vertebrates become increasingly subject to hunting by the influx of people. The extensive trade in wildlife from tropical forests places direct pressure on populations of many vertebrates despite attempted regulation by government-imposed control (Wilkie *et al.* 1998). Illegal harvesting and trade continues unabated principally because wildlife is considered to be common property. The financial burden of conservation is therefore placed on the state and there is little incentive for resource management by self-interested individuals for financial gain through sustainable harvest. Thus conservationists and economists often argue that wildlife resources need to generate income that can be appropriated by private resource owners or managers to create incentives for preservation. However, there are problems associated with commercialization for the purpose of conservation. Resource owners may have difficulty in regulating the harvest of migratory species or of species that forage over areas that exceed single-owner control. Some vertebrate populations may fluctuate widely due to conditions that are beyond the control of the resource manager (e.g. disease or climatic fluctuations), leading to uncertain income generation. There may also be costs in maintaining high densities of wildlife through damage to crops and property (Naughton-Treves 1998; Norton-Griffiths 1998). Furthermore, it is often not clear that sufficient income can be generated from an ecologically sustainable harvest.

A popular alternative to hunting to promote wildlife conservation by income generation is wildlife tourism (Weber 1993; McNeilage 1996). Tourism has the capacity to promote support for conservation by generating employment, increasing foreign exchange earnings and attracting investment capital. For example, gorilla tourism in Democratic Republic of Congo (Zaire) and Rwanda was correlated with an increase in revenue and a decline in poaching, though establishing causal links with the latter is difficult (Butynski & Kalina 1998). It is now apparent that there are many conservation problems associated with nature tourism, which has been linked to the loss of species and habitat degradation (Butler 1991; Duffus 1993). The success of tourism depends on political stability and if this is not assured then tourism can be a highly uncertain source of revenue, as illustrated by the cessation of revenues from gorilla tourism in Rwanda following the civil war in 1994. For conservation benefits to be realized there must be positive economic gains accruing to the people living close to the wildlife populations; however, in most cases revenues flow directly to central government and little is reallocated to the local people. At the same time tourism can have an inflationary effect on local economies. Tourism may also cause changes in the behaviour and fitness of target populations, and the potential risk of disease spreading from humans to habituated primate groups is a major concern that is rarely addressed (Butynski & Kalina 1998).

A community-based approach that advocates giving people an economic stake to ensure sustainable management of their wildlife resources is epitomized by the CAMPFIRE programme in Zimbabwe (Metcalfe 1994). Focusing on communal lands it provides local people with the authority to manage wildlife (Cumming 1990). The principles behind communal wildlife management are elaborated by Murphree (1991). Central to the debate are issues of rights and responsibilities of those who stand to benefit from the commercial exploitation of wildlife. Zimbabwe has gone some way towards exploring the hierarchy of accountability arising from the integration of local interests with those at regional and national levels.

Logging, habitat degradation and conversion are likely to be the primary causes of vertebrate declines and extinctions for some time to come, but as forests become accessible to ever-greater numbers of people the pressures of hunting and tourism will become increasingly important. Privately regulated hunting and tourism are unlikely to ensure the long-term sustainability of vertebrate populations. Care has to be taken that economic incentives are not promoted to the extent where the conservation principle is forgotten. Rather, integration of economic gain with conservation objectives requires that revenue be obtained in the context of a conservation-orientated policy framework based on objective ecological science.

## CONCLUSIONS

Forest managers face a difficult task in managing forests for both timber production and wildlife conservation. Information is lacking on critical issues such as the distribution and ecology of component species, the contribution of biodiversity to the resilience and resistance of ecosystems, and the appropriate spatial and temporal scales to which management must be directed. Even monitoring biological diversity to assess the success of management strategies is subject to difficulties, reflected in the lack of universally acceptable monitoring protocols. Objectives defining desired production outputs and conservation priorities need to be more clearly defined, and terms such as 'biodiversity', 'ecosystem health' and 'ecosystem sustainability' are not amenable to universal definition and too broad to be useful in evaluating management strategies. A more effective strategy for forest management, which combines the twin aims of production and conservation, is to focus on conservation strategies for species that contribute to the maintenance of ecological interactions. Thus it may be more effective to focus on species groups known to have a large impact on community structure and dynamics, i.e. the 'ecosystem engineers' and 'keystone species', as well as the suite of organisms that have important roles as pollinators, dispersers, decomposers, etc. Management for the maintenance of these species has to consider all scales at which the species operate, defined by their resource requirements, mobility, body size, metabolic rates and life-history and demographic attributes (Hansen & Urban 1992). Evaluation of the extent of ecological redundancy (Walker 1992) in forest systems will help to set conservation targets that are integrated with production objectives.

Forest managers must proceed with management activities despite limited availability of ecological information. Spatiotemporal attributes, such as topography, seasonality, roads and water courses, and the natural disturbance regime affect both land-use decisions and habitat use by animals. In the short term, managers must make more effective use of the spatiotemporal information available and, particularly, need to recognize the importance of planning at several nested spatial and temporal scales. Likewise, knowledge of the ecological processes affecting community composition, productivity and turnover is needed to predict environmental responses to disturbances. In the longer term, field studies and monitoring programmes implemented by managers will provide this information to be used for evaluation and improved adaptive forest management, and the costs involved in undertaking such research is likely to be offset by the social, economic and ecological benefits gained.

# 6

# Biological Interactions and Disturbance: Invertebrates

## DESCRIPTION OF PROCESSES OR CHARACTERISTICS

### Ecological attributes that determine the diversity and abundance of invertebrate species in forest ecosystems

This introductory section deals with basic ecological principles that apply in most forest ecosystems, regardless of the region in which they are found.

The diversity and abundance of invertebrate species in forest ecosystems are determined by the interaction of complex ecological and environmental factors. Research into these factors is gradually providing data that can be incorporated into models, both descriptive and mathematical, of the processes involved. At the core of such models is the description of the availability of resources that enables a particular invertebrate species to reproduce, whereas the rate of reproduction is dependent on multiple factors, including temperature, competition, etc., that determine utilization of available resources. In relation to exploitation of resources within a forest ecosystem, the primary focus for forest managers is the potential of invertebrates to cause damage and to reduce yields of the trees themselves. This is certainly the case in managed forests where the primary aim is to grow trees for commercial gain, often with associated value from recreational, landscape or environmental considerations. Any invertebrates that feed directly on the trees, or have an influence of how the trees grow, may be classified as pests and hence may require some form of pest management to reduce the threat (Speight & Wainhouse 1989). Methods of pest management are discussed in Volume 2, Chapter 8, but it is pertinent to consider how ecological theory and practice can be used both to understand why a pest problem has arisen and to provide guidance on how to combat that problem. The same principles can be applied to understanding and enhancing the biodiversity of forests, including conservation of rare and beneficial invertebrates. Indeed, the two extremes of pest outbreak and invertebrate rarity are part of the same continuum of interaction between invertebrates and their host trees.

In general, considering that trees are known to harbour the greatest numbers of insect species (diversity) (Crawley 1983; Stork *et al.* 1997a), it is surprising that serious pest status is reached only relatively infrequently. For example, although oak (*Quercus* spp.) is known to support at least 450 species of insect in Britain (Kennedy & South-wood 1984), Bevan (1987) lists only 35 species as pests, of which five were classified as important or severe (XXX-XXXXX on Bevan's scale). Wider listing of pest species has been noted for forest insects in continental Europe, although even here the number of consistently serious pests is small relative to the total diversity of insects present (Klimetzek 1993).

Naturally, the severity of a pest outbreak will be classified differently depending on the purposes of the forest manager, and thus it is not always possible to classify insects consistently as pests across geographical ranges or, indeed, between years. Further, an insect may normally be classified as uncommon or rare but, given an abundance of resources, may be described later as a pest. This was certainly the case for oak pinhole borer, *Platypus cylindrus* (Coleoptera: Platypodidae), which prior to the 1987 great storm in southern England was listed in the British Red Data Book as rare (Shirt 1987).

Winter (1988) warned that the great increase in availability of freshly dead oak following the storm could result in an increase in numbers of *P. cylindrus* and, subsequently, it has been noted that the insect is now causing damage to high-value oak logs as a result of its habit of boring directly into the heartwood. Therefore, consideration of why invertebrate abundance might increase above a given threshold level can provide a more consistent appraisal of status and also potential approaches for reducing populations below that threshold or indicating measures that can be taken to encourage rare or endangered species.

Among the many interacting processes that determine insect diversity and abundance, a few attributes that are particularly well linked to the environmental status of trees can be distinguished. These provide a means to describe intrinsic processes but also allow comparison with other plants, thus offering the potential for a more consistent approach to descriptive population ecology of invertebrates irrespective of the host plant being considered (Strong *et al.* 1984). Some of these factors are described below.

### *Longevity*

Trees are the longest-lived components of an ecosystem and, in the majority of natural situations, approach or represent the climax stage of vegetation succession. This provides opportunities for continuity of insect population growth and over time this will give a particular set of insect associates for a given tree species (South-wood *et al.* 1979). Continuing the theme of resource availability, the fact that trees may be present in the same location for tens or hundreds of years ensures that both migrant and, particularly, relatively sedentary invertebrates can exploit them at some stage during their growth. However, longevity *per se* is not a quantifiable attribute but does provide the means for the other components, described below, to influence invertebrate diversity and abundance. Increasing stability in the environment, as represented under forest conditions, leads to stability in both microclimate and resource availability, particularly favouring exploitation by specialist insect feeders and their associates.

### *Biogeography*

Tree species richness in a given region or locality varies considerably with geographical location and thus the climax community structure and longevity of individual species may vary accordingly. Although subject to uncertainty and lack of unequivocal evidence of the underlying processes, there is certainly a gradient of species richness of plant cover, especially trees, from the tropics to the poles (Begon *et al.* 1996). Not surprisingly, the same applies to a wide range of animals and plants; as an extreme example, the species richness of orchids ranges from 15 to 2500 in a latitudinal range from 0 to 66° S (Dressler 1985). Further examples are provided by Brown (1988). More specifically, in an analysis of species richness of deciduous forests both in the tropics and, latitudinally, within North America, Brown and Gibson (1983) contrasted the high diversity of tree species in the tropics (up to 100 $ha^{-1}$) with the low diversity from north to south within North America (1–30$ha^{-1}$). However, even within such a generalization there are exceptions, so that for example conifers are more diverse in temperate latitudes.

### *Geological time*

The ecological structure of forests and of their associated invertebrate faunas is determined by both present and recent characteristics, particularly area, and by their history in geological time. The latter characteristic is driven as much by *time per se* as by the history of major disturbances (glaciations, volcanic eruptions, land shifts, etc.) within geological time. Thus, present flora and fauna may well be determined by recolonization since the last major disturbance, the diversity of species being a reflection of both colonization and speciation relative to more recent ecological conditions. The gradient of species from polar to tropical regions could therefore be explained by the high level of disturbance closer to the polar masses compared with the relative stability of climate in the tropics (Brown 1988). Evidence from habitats that have become isolated since the last major disturbance indicates that current species diversity may reflect the historical link to the original major habitat. For example, islands that were once joined by land bridges to the larger land masses of New Guinea have species diversities of birds greater than expected compared with other islands that were never connected (Diamond 1975). However, although local ecological communities can be explained by historical patterns, the gradients over wide geographical scales are not so easily attributed to time as the major determinant. The concept that major disturbances drive change undoubtedly carries weight, but closer analysis of tropical rain forests reveals that they too have been subject to disturbances of various sorts and are, in many cases, much changed over geological time. Based on this knowledge Haffer (1969, 1982) proposed a theory that disturbances in the tropics during the Pleistocene (alternating wet and dry periods giving rise to forest blocks surrounded by grassland) gave rise to refugia in which high levels of speciation took place, ultimately leading to the current high diversity characteristic of the tropics. While this theory is plausible and has been invoked by other authors (Connor 1986), tests of the hypothesis have cast doubt over its wide applicability (Lynch 1988; Flenley 1999).

Irrespective of the mechanistic bases for variation in species richness over wide latitudinal ranges, time is a significant determinant at a more local scale. Birks (1980) analysed the insect faunas of trees in Britain during the period since the last glaciation 13000 years ago and concluded that there was a significant correlation with the time that a given tree species had been continuously present in Britain during that period. Taking this further, Kennedy and Southwood (1984) included Birks' radiocarbon estimates of time in Britain as a variable in a multiple regression analysis of insects on British trees. They showed that time was a significant predictor, both in its own right and as a co-predictor with log abundance (area). This is logical considering that over time there will be a tendency for well-established tree genera to increase their abundance, despite the recent proliferation in areas arising from commercial afforestation.

### *Size*

A consequence of basic structure and longevity is that trees tend to be the largest components of the phylloplane, thus providing a much wider number and range of ecological niches compared with herb and shrub layer plants (Crawley 1983). Such attributes enable more insect species to colonize the plants without facing competition, both interspecific and intraspecific (Strong *et al.* 1984). Size has many interrelated characteristics that help to explain the presence of both herbivores and their natural enemies.

- A larger physical presence is more easily detected and colonized by invertebrates. This concept, termed 'apparency', reflects a higher visual profile for colonizing adults or passively migrating immature life stages (e.g. mean height was a good predictor in models of the distribution of Scolytidae in Finland; Heliovaara & Vaisanen 1995) and a greater production of olfactory cues (e.g. the bark beetle *Tomicus piniperda* can recognize pine tree volatiles and assess suitability for breeding while in flight; Byers *et al.* 1985).

- Complex plant architecture is a consequence of greater size and for trees is particularly linked to the provision of many niches capable of supporting invertebrates with a wide range of feeding strategies (guilds). Although the concept of the guild is not always clearly defined (Hawkins & MacMahon 1989), it is nevertheless a useful means of describing the trophic interactions of invertebrates and trees without having to consider their precise taxonomic status. Thus, seed, bud, leaf, bark and wood feeders can be distinguished, each occupying a different part of the tree and therefore potentially avoiding competition.
- Increased biomass provides greater food resources per individual, thus encouraging growth and breeding success whilst reducing potential intraspecific competition. Price (1992) has reviewed the role of plant resources in insect population dynamics and concludes that, in part, the carrying capacity of an ecosystem depends on plant succession in time and space. Carrying capacity is therefore greater for trees, which generally appear late in succession, compared with herbs or shrubs. Thus biomass is available for longer (see below) and in greater quantities on trees compared with other plants. • The concept of enemy-free space applies to the likelihood of herbivores being overlooked by searching parasitoids or predators (Jeffries & Lawton 1984). One aspect of size is that there is more opportunity for herbivores to occupy a part of the plant that offers protection against natural enemies (and competition from other herbivores). The complexity and number of niches present on a plant (plant architecture; Lawton 1983) therefore has an effect on enemy-free space. The greater the architectural complexity, the greater the time that natural enemies may have to spend in searching for prey, thus reducing their overall effectiveness and enhancing the survival probability of herbivores.

*Area*

Forests tend to occupy greater areas than other plants. There is evidence to support the view that the greater the area occupied by the host plant, the greater the number of insect species associated with that host. This is called the species-area relationship and was originally linked to the island biogeography theories of species isolation and colonization (MacArthur & Wilson 1967). Species-area relationships have been demonstrated for many plants and animals, including both highly mobile and sedentary species. With regard to the numbers of insect species on trees, data on the British insect fauna, one of the best characterized in the world, have been analysed by a number of authors.

Arising from earlier tentative analyses (Southwood 1961), Kennedy and Southwood (1984) analysed the relationships between insect species and characteristics of tree species in Britain. They concluded that there was a significant relationship between the area occupied by British tree species and the numbers of insect species associated with them.

The area occupancy data were based on county and subcounty information of plant records and were thus potentially overestimates of actual areas occupied. Claridge and Evans (1990) reanalysed the species-area relationships using the same insect data as Kennedy and Southwood but employing actual area information derived from Forestry Commission census data (Locke 1987). They concluded that area was not a good descriptor for insect species on deciduous trees but was just significant for conifers.

However, further unpublished analysis of the data broken down into numbers and areas of habitat patches occupied by the tree genera indicated that whereas total area was not a good descriptor, the degree of fragmentation of that area provided a better explanation for the observed insect species numbers. A similar conclusion, concentrating on Lepidoptera that have moved hosts to include conifers, was reached by Fraser and Lawton (1994) in an analysis of new associations on British trees. The overall finding that area provides a broad descriptor of insect diversity remains valid, although this does not explain the mechanisms that lead to these associations.

*Taxonomic relatedness*

In a given ecosystem, the closer the taxonomic relatedness of host plants, the more similar the numbers of insect species associated with those plants. This has been documented for tree species that have a long history in a particular location and thus provide rich sources of invertebrate species that may subsequently colonize other tree species, especially if they are taxonomically related (Claridge & Wilson 1981; Kennedy & Southwood 1984). A particularly good example of the process in action is the colonization of *Nothofagus*, a member of the Fagaceae newly introduced to Britain, which has been colonized rapidly by insects associated with closely related members of the native Fagaceae (Welch & Greatorex-Davies 1993).

Geographical isolation as a critical factor in determining invertebrate diversity and implications for international movement of pest organisms

Geographic isolation drives the acceleration of speciation and provides the means to derive principles on the processes involved. In addition, the fact of isolation means that some invertebrates abundant in one region will be absent elsewhere, despite conditions being suitable for their establishment. In such situations, it is pertinent to examine the international plant health implications of geographical isolation and how modern patterns of trade influence the degree of isolation and later colonization by new insect species.

*Geographical isolation as a factor in determining insect diversity*

In general, the basic principles apply across ecosystem boundaries and geographical zones extending from the poles to the tropics. In addition, the isolation afforded by relatively small islands provides further evidence of the principles of species colonization over space and time. Great Britain provides a particularly good example, because of its island position in the Atlantic Offshore climatic zone and the fact that it has been subjected to glaciation in recent geological time. Among the ecological characteristics that arise from this set of circumstances are a relatively depauperate mix of both animal and plant species that have arisen from the combination of losses resulting from the last ice age and the slow rate of recolonization arising from island status. British flora and fauna are therefore impoverished relative to continental Europe (Rackham 1990). Further influences arise from the preponderance of exotic tree species in commercial forestry in Britain, which although dominated by exotic conifer species also has increasing numbers of exotic broadleaved species (Locke 1987).

Expanding on the principle that colonization of tree species is determined by the features described in section 6.1, the ranges of numbers of insect species associated with trees in Britain reflect both the area (with provisos) and length of time that trees have been continuously present in Britain (Kennedy & Southwood 1984). However, there are exceptions, so that Sitka spruce in particular would appear to support far fewer species than predicted on the basis of area, although this probably also reflects the relatively short history of the species and recent expansions in areas planted in Britain.

### *Implications for international movement of pest insects*

Trading patterns between countries have had a considerable influence on both the flora and fauna of most of the countries in the world. This reflects deliberate introductions of plants and animals for agricultural, forestry and amenity purposes and also accidental introductions by association with other commodities. In the majority of cases, the introduced species have had a neutral to beneficial effect, reflecting their 'domesticity' within managed crop or animal husbandry systems. However, there are many exceptions to this generalization and in most cases the results of introduction of new species were not foreseen, either through lack of knowledge of the potential impacts or because the introduction was accidental and undetected. The literature on international biological control is dominated by examples where attempts have been made to rectify the effects of new pest introductions (both animal and plant) (Waage & Greathead 1988). A classic example is the control of the woodwasp *Sirex noctilio* (Hymenoptera: Siricidae) in exotic pine forests in Australia with the introduced nematode *Deladenus siricidicola* (Nematoda: Neotylenchidae) isolated from *Sirex* in New Zealand. This nematode has a life cycle that includes a free-living form that feeds on the fungus

*Amylostereum areolatum* and a parasitic form that invades *S. noctilio* larvae and subsequently migrates to the reproductive organs of the female wasp, causing sterility (Madden 1988). In this situation, all elements of the interaction are exotic to the new location, with little prospect of natural control being realized.

## PRINCIPLES AND DYNAMICS THAT GOVERN EVENTS AND OCCURRENCE

Expanding on the basic principles in the introductory section, the remainder of this chapter deals with more specific consideration of dynamic processes. This is illustrated by reference to factors that determine the interface between endemic and epidemic insect populations in forests and woodlands. The emphasis is on feedback processes that tend to keep insect populations 'in balance' with their food sources, particularly taking account of the attributes. Departure from this 'balance' can result in either development of pest status or reductions in populations of rare insects, possibly leading to extinctions.

### Interactions with the food plant: phytophagous invertebrates (herbivores)

It has already been mentioned that insect densities generally do not reach damaging levels, thus implying that there are constraints to population growth. Important determinants of population growth are palatability and nutritional suitability of the host tree. Both adult and larval stages may be affected by these parameters. In the majority of cases, adult insects locate suitable host trees for their progeny and lay eggs accordingly. The evolution of oviposition behaviour and host preferences in Lepidoptera has been discussed by Thompson and Pellmyr (1991), who considered host chemistry, plant distribution and suitability of resources as well as the evolution of the wingless forms of many lepidopterous species. They concluded that no single factor governs the evolution of host specificity and that many studies of the mechanisms of selection fail to separate the many possible host selection processes. In the case of flying females, selection of oviposition sites by females is determined by choice, either directed by the cues given off by the host plant, or by landing and 'tasting' the substrate for suitability. An interesting example of such a phenomenon is the teak defoliator moth *Hyblaea puera* (Lepidoptera: Hyblaeidae) in tropical teak forests, especially in India, which swarms as both males and females and then disperses *en masse* to eventually locate and oviposit in the crowns of teak trees that may be considerable distances from the site of origin. Considering that in its natural habitat teak is a relatively localized tree species, growing within mixed forests with many other tree species present, this implies that the moth has well-developed host location capabilities even when the tree is intimately mixed with other tree genera (Nair 1988). Once the tree is located, eggs are laid on the expanding 'tender' leaves, thus providing the newly hatched larvae with the only food suitable within the teak canopy. This and other examples indicate that active host choice by female invertebrates is often linked to the suitability of the oviposition site, or its immediate vicinity, for consumption by the more vulnerable first instar larvae (Trewhella *et al.* 1997). By contrast, those insects having flightless females have more limited opportunities to place their eggs in suitable locations on the 'ideal' host plant. Fraser and Lawton (1994) considered this phenomenon in assessing expansions in the host ranges of British moths and concluded that a greater than expected proportion of 'host-shifting' moths were unable to fly actively to place their eggs. Moths with flightless females, such as winter moth *Operophtera brumata* (Geometridae), European gypsy moth *Lymantria dispar* (Lymantriidae), fall cankerworm *Alsophila pometaria* (Psychidae) and wattle bagworm *Kotochalia junodi* (Psychidae), tend to lay eggs either close to the pupation site (*O. brumata* and *K. junodi)* or more indiscriminately on both the original plant and on other substrates in the vicinity *(L. dispar).* Dispersal in these moths tends to be through the action of undirected larval movements, often by wind carriage of early instar larvae on threads of silk. Interestingly, Hunter (1991) included flightless females among the characteristics inherent in outbreak species of Lepidoptera, although other features such as polyphagy and high fecundity played equally important roles.

A further aspect of host selection and success in establishment is the coincidence of egg hatch, which determines when the vulnerable first larval stages commence feeding, and host phenology. This is particularly crucial for spring-feeding larvae that rely on newly expanding buds and shoots for initial establishment of young larvae on the most succulent tissues. This is most apparent in those species of insect that overwinter in the egg stage and where early instar larvae must feed on newly expanding foliage in the spring to survive. Among such species, oak leaf roller moth *Tortrix viridana* (Lepidoptera: Tortri-cidae) is particularly vulnerable to lack of synchrony between egg hatch and budburst of oak. Studies in France have demonstrated that populations of the moth adapted to specific budburst periods of *Quercus pubescens* (early budburst) and *Quercus ilex* (budburst up to 3 weeks later) could be distinguished in terms of egg hatch and hence early larval establishment (Merle 1983). Hunter (1992) also confirmed the dependent relationship between oak budburst phenology and establishment success of neonate *T. viridana* and *O. brumata* larvae. Similar results were obtained in tests of clonal poplars in relation to timing of budburst and foliar suitability (nutrition and phenolic defences) for larvae of forest tent caterpillar *Malacosoma disstria* (Lymantriidae) (Robison & Raffa 1997).

The development of tissues with differing degrees of palatability to potential herbivores is an intrinsic part of phenological coincidence between herbivore and host. Palatability may vary widely as plant parts develop during the growth season and also between seasons and is linked to both physical structure and chemical composition of the different tissues of the host plant (Price 1992). For example, the suitability of oak foliage to a range of herbivores is determined in part by reactions to the presence of tannins (negative) (Feeny 1976), total phenolics (negative) (Rossiter *et al.* 1988), foliar astringency (positive) (Hunter 1997) and nutritional content (Hunter & Schultz 1995).

In general, there is a feedback relationship between the degree of defoliation suffered by a tree and the absolute and relative quantities of phenolic and nutritional components present in the leaves. This apparent trade-off between growth and defence is the subject of considerable research and provides some interesting insights into how plant quality can act on population dynamics of herbivores. It is also an intrinsic component of multitrophic interactions in which other factors, particularly the roles of natural enemies, affect the dynamics of herbivores.

Plants that are long-lived, such as trees, could be expected to develop more effective defence mechanisms than transient plants because of the longer time period over which they would be subjected to potential herbivore attack and their greater apparency in the environment (Rhoades & Cates 1976). While it is true to say that trees do have a wide range of defensive strategies, it has not been possible to confirm that overall they are better defended than less apparent plants. The hypothesis was tested specifically by Reader and Southwood (1981) who used generalist herbivores to assess the palatability of plants with turnover rates ranging from rapid to infrequent, and showed an overall decrease in palatability with increases in successional stage, i.e. towards longer-lived plants. However, results were very variable and indicated that there were clearly a wide range of factors involved, plant chemistry being one of the more important.

The mechanistic basis of plant resistance is not universally agreed and indeed it is apparent that the actual mechanism involved will differ, depending on a combination of host plant genotype, environment and herbivore pressure. At its most simplistic level, plant defences involve both the presence of chemicals that deter herbivores and the presence of nutrients that enable them to grow and ultimately reproduce. An attractive theory proposed by several authors is that plant growth is a trade-off between carbon and nitrogen such that if nitrogen is limiting, then plant growth slows down, leading to excess carbon being diverted to secondary chemicals and thus improved plant defence (Bryant *et al.* 1983, 1993). Ruohomäki *et al.* (1996) questioned this assumption in their studies of nitrogen fertilization and its effects on mountain birch. They concluded that a resource-based hypothesis did not provide an adequate explanation for observed events and that a source-sink hypothesis was a better descriptor. The latter hypothesis assumes that the plant is split into a number of interlinked components, some of which act as sources of nutrients and carbon (e.g. photosynthesizing leaves act as sources for amino acids and provide the necessary components for future resource allocation to buds) and others as sinks (e.g. newly expanding buds take nutrition from a general 'pool' in the tree). Haukioja and Honkanen (1996, 1997) analysed various forms of induced defensive reactions in relation to the source-sink hypothesis. For birches and pines, they showed that foliar damage and the immediate and delayed effects on foliage suitability for herbivores could be explained by impacts on physiological sinks, such as newly expanding buds. This had the effect of reducing plant growth in the following year and that impact was local to the site of damage, reflecting the damage to the particular meristems affected (Honkanen & Haukioja 1994). Other studies, notably those of Bryant *et al.* (1993) on paper birch *Betula papyrifera* and Hunter and Schultz (1995) on *Quercus prinus*, showed that addition of nitrogen fertilizers reduced the ability of trees to produce secondary defensive compounds. This apparent contradiction between nutrient balance and source-sink hypotheses may be a result of specific differences in reactions depending on tree species or may represent artefacts of experimental design. For example, is there a threshold effect such that addition of excess nitrogen can override internal feedback mechanisms in the plant? The studies of Bryant *et al.* (1993) may well support the contention that excess nitrogen can mitigate any defensive characteristics of the host plant. Clearly, further work needs to be done to elucidate the precise mechanisms involved.

Irrespective of the mechanisms, there is evidence to indicate that the suite of defences produced by trees can be effective in reducing attack levels by different herbivores. Such defences vary in their effects on different feeding guilds so that it is possible to distinguish between the strategies developed by both the herbivores themselves and the plants on which they feed. This has been reviewed comprehensively by Koricheva *et al.* (1998) who carried out a meta-analysis of published experiments in which the effects of stress were analysed against a series of parameters, including feeding guild, plant phenology, plant structure, etc. The authors concluded that, overall, plant stress had little effect on insect relative growth rate, fecundity, survival and ability to colonize a plant. However, when data were analysed at the level of the insect feeding guild, there were large and significant differences in responses, both positive and negative, to plant stress (Table 6.1). Thus, insects that colonize trees by boring into the bark and/or wood (bark beetles, longhorn beetles, woodwasps, etc.) were more successful in colonizing stressed trees (there was no information on the other parameters, but it would be reasonable to assume that successful colonization leads to a higher probability of success in the other parameters). Sucking insects, which use stylets to access nutrients in the phloem, also performed better on stressed plants. In contrast, gall-forming and chewing insects tended to be negatively affected, the latter being particularly suppressed on slow-growing plants such as trees.

## Interactions between herbivores and their natural enemies

Discussion has so far been concerned with the trophic interactions between herbivores and their host trees, providing examples of the so-called bottom-up approach to understanding the dynamics of the two elements. However, a further trophic level that interacts with, and may influence critically, herbivore dynamics is the role of biotic mortality factors, collectively termed 'natural enemies'. Such factors can impinge on any stage of the herbivore life cycle and may be linked in a quantitative way to the population size of the host or prey. There is considerable debate concerning the relative importance of top-down impacts of natural enemies compared with the resource-driven, bottom-up effects of the plants themselves (Dempster & Pollard 1986; Harrison & Cappuccino 1995). However, recent literature is tending to take a balanced view that includes both resource and natural enemy impacts in determining the dynamics of herbivore interactions with their host plants (Kidd & Jervis 1997; Schultz 1997). There is no doubt that natural enemies have a role to play in determining the numbers of potential prey within a given ecosystem. Regulation of prey densities within predictable limits requires that the numbers of natural enemies are linked quantitatively to those of their prey (density-dependent regulation) (Hassell & Waage 1984). Although many models have been developed in an attempt to predict the relative dynamics of natural enemies and their prey, the most pertinent evidence in relation to management of forest insect pests comes from their demonstrated use as biological control agents (Waage & Greathead 1988). Although biological control has ranged from highly successful, requiring no further intervention, to only partial, overall the data lend strong support to the supposition that natural enemies can be important regulatory factors in herbivore dynamics. Kidd and Jervis (1997) considered the roles of para-sitoids and predators on forest insect populations in some detail. They concluded that natural enemies were significant in regulation of forest pests but that they could also give rise to population increases as a result of delayed density-dependent processes. Interestingly, Kidd and Jervis considered the bottom-up vs. top-down debate by assessing the dynamics of the natural enemy-herbivore-plant interaction within the context of the synoptic population model proposed by Southwood and Comins (1976). Herbivores were classified into groups according to their growth rates and ability to disperse.

- *r*-strategists have high growth rates and effective dispersal powers, making them capable of colonizing and utilizing unstable or transient habitats. These species tend to have few effective natural enemies and are usually regulated by resource availability.
- ^-strategists represent the opposite extreme, having slow growth rates and limited ability to disperse and are associated with stable habitats. Population size tends to be tightly linked to availability of resources and natural enemy activity has little influence on total numbers.
- *r-K* strategists have some of the characteristics of both groups and tend to lie in the 'natural enemy ravine' where the range of natural enemies is greatest and have pronounced numerical responses to changes in herbivore densities. Forest pests tend to be classified as *r-K* strategists but, because they tend towards the endemic in most cases, will be skewed to the *K* end of the continuum.

Evidence that natural enemies have a significant influence on forest insect numbers comes from examples of successful biological control in many countries of the world and from studies of the dynamics of some of the most damaging forest pests. The Lepidoptera and Homoptera dominate, reflecting the importance of these orders as damaging agents of forest trees, although in virtually all cases there are contributory influences of host plant resources, especially secondary chemicals. The Coleoptera, particularly the bark beetles, tend to have a dynamic interaction with the host plant, which may act in concert with the activities of natural enemies. This is certainly the case for *Dendroctonus micans*, which is regulated by both the action of its specific predator,

*Rhi-zophagus grandis*, and the defensive characteristics of its host spruce trees (Wainhouse & Beech-Garwood 1994; Evans & Fielding 1996).

The characteristics of successful natural enemies include a well-developed ability to locate suitable hosts (non-random search), high reproductive capacity, quantitative linkage to changes in host density and dispersal capacity to exploit spatially separated host populations. These clearly apply to many of the predators and para-sitoids that have been studied but are less easy to apply to pathogenic organisms equally important as mortality factors.

### *Predators*

Invertebrate predators search for and consume their prey directly, usually in both the adult and the larval stages. The best-known examples tend to be among the ladybirds (Coleoptera: Coccinell-idae) familiar to many in urban and woodland situations. However, among the predators of forest pests it is the beetle predators in the families Cleridae, Trogostidae and Rhizophagidae that have been most studied in relation to their impacts on bark beetles (Dahlsten & Berisford 1995). Predators are not restricted to the insects; mammals, birds, reptiles, spiders and mites have all been shown to feed on tree-dwelling herbivores. In most cases the predators are generalists and will feed on the most abundant prey available. However, within this generalization, some apparently polyphagous predators act quite selectively and respond directly to the pheromones (kairomones) produced by their prey. Thus, the clerid beetle predator *Thanasimus undatulus* is attracted to the pheromones of spruce beetle *Dendroctonus rufipermis* and the secondary bark beetles *Dryocoetes affaber* and *Ips tridens* (Poland & Borden 1997). In this case, use of the /. *tridens* pheromone produced a dual effect on the primary pest *D. rufipermis* by both competitively excluding the spruce beetle and by attracting the predator to the tree. Other generalist predators have significant effects at low but not at high densities; the white-footed mouse, *Peromyscus leu-copus*, has been shown to regulate low-density gypsy moth, *L. dispar*, populations through predation on late larval and pupal stages (Campbell & Sloan 1977). However, this effect was not apparent at high gypsy moth densities, presumably because the predator numerical response was insufficient to maintain density-dependent regulation (Campbell & Sloan 1978).

*Parasitoids*

Insects that parasitize their hosts through laying eggs in, on or near host individuals and in which their larval stages develop are known as parasitoids. This group of natural enemies is extremely common and probably makes up over 10% of insect species in the world (Godfray 1994). The majority of parasitoids are in the order Hymenoptera, notably the families Ichneu-monidae and Braconidae. Dipterous parasitoids, in the family Tachinidae, have also been shown to be effective in regulating prey, for example *Cyzenis albicans*, a tachinid parasitoid, is a regulator of winter moth, *O. brumata*, populations in Canada where it was introduced from Europe (Roland & Embree 1995). Interestingly, *C. albicans* is a density-independent mortality factor in winter moth populations in Europe but is strongly dependent on larval density in Canada (Hassell 1980). The interactions between natural enemies, particularly parasitoids, and their hosts have been the subject of much research into the dynamics of the interactions. Mathematical models have been produced that explain some of the variability in responses between parasitoids and their hosts and are being increasingly refined to take account of both temporal and spatial parameters (Pacala *et al.* 1990). Models that take account of dynamics in a single isolated population tend to be unstable and therefore the dynamics in neighbouring populations also need to be taken into account. Comins *et al.* (1992) simulated predator-prey interactions as a series of metapopulations in which both organisms were allowed to disperse between patches. Dynamics of individual patches were unstable but stability was observed in the system as a whole, thus demonstrating that persistence of the whole population of either predator or prey could be achieved despite instability within its subcomponents. Such simulations show that it is necessary to have not only a detailed knowledge of natural enemy behaviour and dynamics but also an appreciation of the spatial elements in the interaction with its prey.

Overall, therefore, there is evidence to show that parasitoids, by virtue of their directed search behaviour and relatively high specificity, can have a regulatory influence on host dynamics. This does not necessarily imply that the host will be regulated below an acceptable economic threshold and thus it may be necessary to take account of other mortality factors to explain fluctuations in the host population.

*Pathogens*

Invertebrate herbivores are subject to a wide range of naturally occurring pathogenic organisms, many of which will not be recognized as such in assessing mortality rates in the field. Pathogens differ from other natural enemies in not having the ability to search actively for their hosts. Dispersal is therefore a passive event that relies on biotic and abiotic factors to ensure that the pathogen is able to encounter suitable hosts. Replication and infection arise only when the infectious unit is ingested (bacteria, Protozoa and viruses) or comes into contact with the integument of the potential host (fungi) (Fuxa & Tanada 1987). Although this introduces a high element of chance in determining encounter frequency, pathogens have a number of strategies that help to overcome this potential problem.

*High reproductive rate.* Replication results in massive increases in the quantity of inoculum produced on death of the infected host. For example, the baculovirus of gypsy moth, *L. dispar*, increases 10000 fold compared with the initial lethal dose to yield over $10^9$ inclusion bodies per larva (Entwistle & Evans 1985). Massive increases of this nature ensure that large quantities of inoculum are released in the immediate area where other hosts are likely to be feeding, thus partially overcoming the inability to search actively for hosts.

*Ability to persist.* Although pathogens are subject to rapid inactivation from environmental factors such as ultraviolet light (Ignoffo 1992), they are still able to persist in the environment between host generations. This arises from a number of pathways, the most significant being the protection of the infectious unit within an outer coat (spores for bacteria, fungi and Protozoa; proteinaceous inclusion bodies for several groups of insect pathogenic viruses) and the tendency for inoculum to be deposited in protected situations such as soil, bark crevices, etc. (Evans & Harrap 1982). Infection in the following generation depends on encounter by early larval stages, which almost invariably are the most susceptible and thus require only very small quantities of inoculum to induce infection. The massive increase in secondary inoculum arising from primary infection can then lead to a local pool of inoculum available for encounter by other individuals in the same host population, thus leading to accelerating levels of infection and further increases in inoculum (Anderson & May 1981).

### Transovarial transmission.

In some cases, such as the Protozoa and some non-occluded viruses, the nucleic acids of the pathogen are incorporated directly within the germplasm of the host insect and thus can persist in a latent form to the next generation of the host (e.g. the picornaviruses; Moore & Eley 1991). However, the evidence for true transovarial transmission within the host cells is often equivocal and difficult to separate from transovum transmission through external contamination with infective stages of the pathogen (Andreadis 1987). In addition, expression of the pathogen in the next generation is an uncertain event and thus in terms of overall dynamics of pathogen infections it would appear that transovarial transmission is a 'failsafe', rather than a primary route of transmission.

### Transmission through other biotic agencies

Many of the pathogen groups listed in Table 6.3 have been shown to be transmitted by biotic agents such as predators or parasitoids. For example, birds eat baculovirus-infected pine beauty moth larvae and void infective virus in their faeces, thus contaminating foliage and inducing infection in other moth larvae (Entwistle *et al.* 1993). Similarly various parasitoids, including Braconidae, Encyrtidae, Ichneu-monidae (all Hymenoptera) and Tachinidae (Diptera), have been shown to transmit bac-uloviruses between hosts (Entwistle & Evans 1985).

Not surprisingly, in view of the debate on bottom-up vs. top-down processes, there is an interaction between the infectivity of various pathogens and the food substrate consumed while the host is ingesting the pathogen. The presence of tannins and other phenolics can reduce the apparent effectiveness of *Bacillus thuringiensis* (Appel & Schultz 1994) and baculoviruses (Foster *et al.* 1992) against gypsy moth on various species of oak tree. In these cases there appears to be significant binding between the inclusion body of the pathogen and the hydrolysable tannins within the oak leaves. Understanding the dynamics of pathogen-host interactions clearly requires in-depth information on the subtle effects at various trophic levels.

## INTERACTIONS WITH OTHER PROCESSES, WITH SCALE AND WITH FOREST TYPE

Many of the processes that govern the dynamics of insect associations with trees were discussed in the previous sections. In the main, these factors combine to keep insect populations at levels that do not appear to pose a threat to the health of trees or their commercial value. As Crawley (1989) has pointed out, there is a common perception that because the terrestrial environment is mainly green, then neither the plants nor their associated herbivores are limited in their resource requirements. However, the previous discussion has shown clearly that there are complex interactions between the different trophic levels associated with trees so that the end-result may give rise to major fluctuations in the relative performances of the different components. In both their natural and managed states, forests are affected by factors of scale such that regional site characteristics and local climate will influence the species mix and the growth rates of trees. Such effects will also influence invertebrate associates, which will respond both to resource availability (at both the plant and higher trophic levels) and to temperature and other abiotic conditions that will drive the rate of response. Some of the influences of scale are implicit in the species-area relationships that, at least in part, provide descriptors of species accumulation of both animals and plants in a particular region (Williamson 1988).

What, then, are the effects of scale on communities of invertebrates in forests and how are they related to forest management in a rapidly changing world? It has already been stated that the area occupied by a particular tree species is an important determinant of insect diversity, particularly if that area contains species that have been present for long periods. Such a situation applies to the natural oak populations in the northwestern USA, where two of the most convincing demonstrations of the species-area effect and interactions with other local and landscape features apply. The natural distribution of oak genera and species in this region covers a wide latitudinal and altitudinal range, including the Californian region (southern Oregon to northern Baja California bounded by the Rocky Mountain and Sierran regions, thus isolating the zone) and Atlantic region (larger, ranging from southern Canada to the Gulf of Mexico, west to the Great Plains). The oak species in the two regions are distinct from each other, those of the Californian region being closely related to Asian rather than eastern US oaks (Cornell & Washburn 1979). Two separate studies of herbivores within specific feeding guilds have been carried out in these regions. Cornell and Washburn (1979) studied the Cynip-inae (Hymenoptera: Cynipidae), a subfamily of gall-forming wasps that are extremely specialized to the genus *Quercus*. They have been extensively studied and the data were used by Cornell and Washburn to investigate relationships to the areas occupied by oaks in the Atlantic and Californian regions. An earlier study by Opler (1974) also assessed relationships to oak distribution but was concerned with leaf-mining Lepi-doptera (Lepidoptera: Gracillariidae) within California only. Both studies demonstrated significant species-area relationships against total range (expressed in square miles). However, the pattern of response was quite variable depending on the subsets of data examined; some of these are reproduced in Table 6.4. The steeper slope and higher coefficient of variation in the Californian subset was attributed to the greater ecosystem variability in the latter location, taking account of the larger number of oak species, the wider altitudinal range and the greater variety of topographical features compared with the Atlantic region (Cornell & Washburn 1979).

Interestingly, the leaf-miner data for the same geographical range of oaks suggest that new moth species colonize sympatric oak species within the same area and that there is no strong link with taxonomic relatedness between oak species, thus demonstrating a response more to total resource availability than to oak species *per se* (Opler 1974). Such data provide support for habitat heterogeneity as one of the principal factors driving species-area relationships. Further studies by Cornell (1986) indicated that both the size of oak trees and, particularly, the size of the pool of cynipine wasp species in a given area were the major determinants of species richness on a given oak species. He attributed this to several of the studies (reviewed in Docherty *et al.* 1997), although the wide variation in responses indicates that the effects may be tree and site specific. Elevated temperatures will also act on performance of both plant and invertebrate, changing both absolute growth rates and, perhaps more importantly, the synchrony between host plant suitability for feeding and presence of the appropriate stage of the insect. Those insects with overwintering stages that require quite precise coincidence of emergence with budburst of their host trees may be particularly vulnerable to changes in the incidence of frosts, especially if emergence is driven more by daylength than by temperature *per se* (Dewar & Watt 1992). Diapause termination, triggered by a specific period of chilling below a subzero threshold temperature, may also be affected so that emergence in the spring may be reduced, thus affecting overall population size. However, some insects presently constrained by low winter temperatures may show enhanced survival and increased reproduction with an increase in warmer winters. This is likely to be the case for green spruce aphid, *Ela-tobium abietinum*, which is normally killed by winter temperatures below -8°C (Straw 1995).

Pollutants, such as ozone, $SO_2$ and $NO_x$, may have direct effects on tree performance and both indirect and direct effects on insect growth (Brown 1995). Ozone results in amino acid changes within trees that can provide a richer food source for phloem-feeding insects in particular, potentially leading to higher growth and reproductive rates (Bolsinger & Fluckiger 1989). However, recent studies by Holopainen *et al.* (1997) indicated that ozone episodes were less important than increased levels of nitrogen fertilizers in determining the performance of the aphids *Schizolachnus pineti* and *Cinara pinea* on Scots pine and *Cinara pilicornis* on Norway spruce. These authors concluded that occasional episodes of ozone may increase amino acid availability but, generally, the effects from predicted increases in atmospheric ozone were unlikely to be significant. In considering effects on herbivores as a whole, several studies have demonstrated the high sensitivity of sap-sucking insects to the effects of atmospheric pollutants such that elevated levels of $SO_2$ and $NO_x$ give rise to greater RGR and hence reproductive success, for example *E. abietinum* responded to increased concentrations of $SO_2$ with increased RGR when exposed on Sitka spruce (Warrington & Whittaker 1990). Foliage-chewing insects have also been shown to respond to atmospheric pollutants but with rather inconsistent trends. Scots pine seedling trees subjected to elevated $SO_2$ had higher sulphur, decreased calcium and higher levels of free amino acids and were shown to reduce the larval performance of pine looper moth, *Bupalus piniaria* (Lepidoptera: Geometridae), but had no effect on pine lappet moth, *Dendrolimus pini* (Lepidoptera: Lasiocampidae) (Katzel & Moller 1993). The authors attributed the different responses to major reductions in glucose and fructose, which adversely affected *B. piniaria* whereas *D. pini* was able to tolerate the lower carbohydrate levels. The foliage of mountain birch, *Betula pubescens* ssp. *czercpanova*, subjected to experimentally enhanced levels of sulphur and nitrogen over a 10-year period had no effect on autumnal moth, *Epirrita autumnata* (Lepidoptera: Geometridae), feeding on the foliage (Suomela & Neuvonen 1997).

Overall, despite high variability, it would appear that atmospheric pollutants have a neutral to positive effect on insect performance, although there are interactions with temperature, nutrient levels, etc. that can induce different effects even within the same insect species. Even when insect performance is impaired (e.g. elevated $CO_2$), the net effect may be increased damage as feeding rate increases to compensate for reduced food quality. This complex area of herbivore-plant dynamics clearly requires further work to determine patterns and causes.

## FACTORS INFLUENCING PROCESSES AND OPPORTUNITIES FOR MANIPULATION

The array of processes considered in the early part of this chapter range from those reflecting different niches within a single tree, through local scales to the ultimate position of trees as major landscape components. It is also apparent that there are complex multitrophic interactions between trees and their herbivores that make it difficult to generalize concerning potential to manipulate processes. Management at the tree or stand level will involve local processes that reduce some of the complexity, such that gross climate effects may be removed as driving variables, although even here there may be microclimate effects at very small scales. Effects at this scale are manifested through the intrinsic suitability of the local population of trees, thus bringing in structural, nutritional and defensive qualities (see section 6.2.1) as well as the local impacts of natural enemies (see section 6.2.2).

The interactions with wider-scale attributes are illustrated schematically in Fig. 6.4. Extrinsic effects act at a range of scales, from local (within the forest), through regional to geographical, such that the effects on a given tree population become less but the wider attributes of climate, topography and site characteristics come into play. This will have the effect of changing the absolute and relative abundances of tree species and hence their associated invertebrate species over wide areas.

The examples of leaf-miners and gall-wasps from the full latitudinal range of oak species in California serve to illustrate this principle. In summary, therefore, any population of herbivores associated with trees is actually influenced by the herbivore-tree interface so that, provided the herbivore is able to find the host tree, its ultimate performance is driven by local factors that may in turn be influenced by the interaction parameters. Management at this local scale will reflect the initial components of tree species choice and location (i.e. the suitability of the site for tree growth and ultimate aim of the crop).

This, of course, applies particularly to artificial forests rather than natural forests, where tree species mix and stage are entirely driven by site and local climate characteristics. Once trees are established, the elements of further management that can influence susceptibility to herbivore colonization are addition of fertilizers, particularly nitrogen, and thinning regimes that will affect proximity to other trees and plants and generate open spaces and habitat corridors. All other factors will be extrinsic (proximity to other forests, topography, climate, etc.), with decreasing influence with both distance and time.

The influence of given factors is therefore a question of scale. If these are at the landscape scale, then it is possible to view forest management at a more remote level so that a high degree of fragmentation at a number of area scales can be considered and thus their relative influences on local herbivore-tree interactions measured at any fixed point within the wider landscape. This conceptual approach is now being used to increase understanding of the wider processes affecting invertebrate interactions with trees, making use of geographic information systems (GIS), new geostatistical methods and improved invertebrate sampling regimes to assess populations over relatively wide areas (see Stork *et al.* 1997a for a number of review articles focusing on canopy arthropods, especially in relation to tropical communities).

Further impetus to such approaches has been provided by the various biodiversity conventions (UNCED, Rio, 1992; Ministerial Conference on the Protection of European Forests, Helsinki, 1993) and the implementation of biodiversity action plans at individual country levels (e.g. Sustainable Forestry, The UK Programme, delivered to Parliament in January 1994). The principles described in this chapter apply equally to conservation of invertebrate diversity and abundance, thus embracing enhancement of biodiversity and avoidance of invertebrate outbreaks within the same conceptual framework.

## Forest fragmentation at local and landscape scales

Among the parameters, the degree of fragmentation of a forest is one of the key variables that can be subject to human influence, both deliberate and accidental. It is also one of the variables most subject to natural disturbances over recent and geological time-scales. There is increasing interest in fragmentation in relation to tropical forests because of the extremely rapid loss of habitat through intensive logging. For example, models have been constructed to take account of the degree of fragmentation of tropical forests in relation to core area, edge effects and the shape of the forest area (Laurance & Yensen 1991). Further refinement, including provision of spatiotem-poral components and the degree of connectivity between fragments, has been provided by Didham (1997). This synopsis included:

- area effects, taking account of total area and interior core area which is not influenced by any edge effects;
- edge effects, especially the distance over which the edge has an influence relative to the core;
- shape of the fragment, which clearly affects both edge and core elements;
- degree of spatiotemporal isolation so that the time since fragmentation is taken into account;
- degree of habitat connectivity in the landscape, thus providing corridors for movement of species through otherwise unsuitable habitats.

Studies of the relative influences of edge to core in forests have been carried out in both temperate and tropical forest ecosystems, using a range of sampling methods. In general, it has been noted that invertebrate densities, of both individuals and species, tend to decline with distance from a forest edge to the core. Thus various beetle families show greater species richness or abundance at the edges of forests: Carabidae (Duelli *et al.* 1990), Cerambycidae (McCann & Harman 1990), Staphylinidae (Buse & Good 1993) and mixed families of beetles (Baldi & Kisbenedek 1994). Although not sampled in relation to direct interactions with host trees, the bark beetle *Ips typographus* tends to colonize edges more frequently than forest interiors (Sanders 1987). Patterns of abundance from edges to core woodland have been assumed to be a smooth gradation but recent studies on arthropod communities in Amazonian forests by Didham (1997) have demonstrated a bimodal form to relationships of edge to core abundances of leaf-litter invertebrates. This effect was compounded with the area of the forest fragment being studied, leading to three-dimensional relationships between fragment area, distance from forest edge and abundance of selected beetle species. Within the Amazonian forest system, the effect of fragmentation was felt even within large remnant forest blocks so that an area considerably greater than 100ha, possibly approaching 1000ha, was required before the invertebrate fauna matched that of undisturbed forests (Didham 1997).

The implications of results such as these are difficult to extrapolate to the heavily managed forest landscapes of Europe and, increasingly, North America and elsewhere. Forest fragmentation can now be regarded as the norm in such situations and therefore the invertebrate faunas in such fragments may already be greatly impoverished compared with those in original undisturbed forests. However, the general principles of maintaining the ratio of edge to core to ensure that the habitat requirements of forest specialists as well as edge colonists are maintained is a sound strategy. Unfortunately, the biological attributes of edges and cores are not clearly defined and thus the type of edge may be as important as the edge itself. For example, studies by Ozanne *et al.* (1997) in canopies of managed conifer forests in Britain showed a lower species richness of various invertebrate taxa at the edges compared with the cores of forests studied. This apparent contradiction to the studies cited earlier may reflect the type of edge, which in managed conifer plantations is often a very abrupt transition from the surrounding open ground area. The authors suggested that a gradation of the edge, with scrubby transition zones, could give rise to greater species richness at the edges but this would probably be at the expense of forest canopy specialists. Such uncertainty reinforces the need for more detailed studies of the mechanistic basis for observed effects of forest fragmentation, particularly taking account of the wide range of ecological factors that have been discussed in this chapter. Studies such as those of Waltz and Whitham (1997) provide useful data on the characteristics of individual trees that may influence edge and core populations of particular herbivores. They studied the effects of canopy structure in clonal poplars in relation to arthropod communities dominated by the gall-forming aphid *Pemphigus betae* and the leaf-feeding beetle *Chrysomela confluens*. There were direct correlations between the mature ramets of the trees and invertebrate species richness and this varied with the clone studied. The authors concluded that such within-tree variability in resource availability explained the higher biodiversity in mixed-aged stands and at edges compared with interiors of even-aged stands.

As Stork *et al.* (1997b) have pointed out, the understanding of forest arthropod dynamics, especially canopy-dwelling arthropods, is hampered by the tendency to gather observational data rather than by rigorous hypothesis testing. They also correctly point out the sheer difficulty of carrying out such studies, graphically illustrated by the 2-year effort by over 20 taxonomists at the Natural History Museum, London to sort and partially identify samples collected over a mere 12-day sampling period in Borneo. The underlying ecological principles that govern invertebrate communities of forests are clearly complex and yet unless they are incorporated into studies of forest fragmentation, there is little prospect of developing a sound science of both pest management and invertebrate conservation for the future.

## Forest disturbance as a factor in invertebrate dynamics

The processes described above emphasize that size and scale matter in relation to communities of forest invertebrates. It was also stated that the areas of natural forest remnants are progressively shrinking and this is likely to have an impact on associated invertebrates. Much of the disturbance that leads to this extensive fragmentation is a result of human activity, directly or indirectly (Vitousek 1994). Anthropogenic influences such as aerial pollution and climate change are having an increasingly global effect that will take many years to influence trees and their invertebrate communities. Prediction of these effects is imprecise because of the relatively poor understanding of the processes involved and the difficulties, mentioned above, of tracking changes against an imperfect baseline.

Some indication of the effects of habitat fragmentation have been provided in studies of the outbreaks of some of the more damaging forest pests and these serve to illustrate the complexity of managing change. Edge effects are well known in relation to initial infestations of several damaging bark beetles, particularly in relation to felling programmes that can provide breeding resources as well as opening the remaining forest to the risks of windthrow, etc.

*Ips typographies*, the most damaging bark beetle in western Europe, is sustained at low densities in most forest locations by reliance on sporadic availability of weakened trees for initial breeding. However, if a sudden increase in breeding material is brought about by felling programmes or wind damage, populations can build up rapidly and attack surrounding living trees

(Christiansen & Bakke 1988). Such eruptive patterns of attack are common in different species of bark beetle and the triggering of the infestation is frequently linked to large-scale provision of suitable breeding material on-site, either through human intervention or via natural effects such as fire or severe winds.

A good example of such an effect was the major fire in Yellowstone Park, USA during 1988 which left a mosaic of fire-killed, fire-damaged and undamaged areas in the area (Rasmussen *et al.* 1996). Surveys showed that 12.7% of Douglas fir *(Pseudotsuga menziesii)*, 17.9% of lodgepole pine *(Pinus contorta)*, 6.6% of Engelmann spruce *(Picea engelmannii)*, 7.5% of subalpine fir *(Abies lasiocarpa)* and 2.8% of whitebark pine *(Pinus albicaulis)* were killed following the build-up of bark- and wood-boring beetles in fire-damaged and stressed trees.

The unexpected and severe nature of the Yellowstone fires also illustrates one of the adverse consequences of fire-management regimes that have been in place in the USA since the early 1900s. Fire was previously a normal component of forest ecosystems in the USA and Canada and had a considerable influence on species composition, succession and forest stability (McCullough *et al.* 1998). Regimes of fire suppression have changed the structure of many forests and, in some cases, have led to insect outbreaks.

A particularly good example of such an effect is the balsam fir and white spruce forest ecosystem of eastern Canada and the north-eastern USA. Fire suppression was started in 1920 and has been accompanied by increased frequency, intensity and scale of spruce budworm, *Choristoneura fumiferana* (Lepidoptera: Tortricidae), outbreaks (Blais 1983).

Apart from any direct effects on the budworm, fire suppression altered succession by reducing pioneer species such as aspen, birch, jack pine and black spruce, thus decreasing the species mix that previously broke up the highly susceptible fir and spruce-fir stands. Recent strategies have been to return to selective use of prescribed burns to help restore the previous forest mosaics and thus to help combat spruce budworm outbreaks (McRae 1994). McCullough *et al.* (1998) cite several other examples of linkage of insect outbreaks to fire suppression regimes and conclude that an improved understanding of the ecology of fire-tree-insect interactions will help to identify strategies to manage pests and to maintain biodiversity in northern forest ecosystems.

## UNDERSTANDING THE FOREST AS A BASIS FOR MANAGEMENT

There have been few published attempts to bring together the various elements that have formed the basis of this chapter in developing rational forest management regimes to provide a balance between pest infestations and enhancement of biodiversity. The need for such balanced approaches is now acute as multipurpose forestry becomes the norm in both developed and developing countries. The interactions summarized in Fig. 6.4 serve to emphasize that it is not sufficient to deal with the relative simplicity of the forest block or stand and that a perception of the effects of landscape scales must also be considered. It is fortunate that the tools to allow such approaches to be attempted are now available and are becoming increasingly sophisticated. GIS, often combined with global positioning systems (GPS), and a range of remote sensing techniques have revolutionized the ways in which landscape and geographical features can be incorporated into spatially explicit models of insect-tree interactions (Liebhold *et al.* 1993). The STS (Slow The Spread) process for gypsy moth in the USA (see Volume 2, Chapter 8) provides a good example of the use of spatiotemporal modelling to assess and predict the effects of management regimes on the spread of the pest into new areas (Sharov & Liebhold 1998). Host plant availability and rates of natural spread as well as incorporation of landscape features (Liebhold *et al.* 1994) have enabled refinements in prediction and management for the gypsy moth STS process. Optimization has led to reduction in the costs of the programme by lowering the numbers of pheromone traps and associated control measures at increasing distances from the advancing front.

Aspects of stand and forest management approaches can also be applied to conservation strategies. The basic tenets of interactions between insects and their forest environment remain the same, although the aims will be very different and will require a balanced approach in order to avoid creating conditions for insect outbreaks while attempting to enhance a particular species or mix of invertebrate species. Such problems are particularly acute in management of old-growth forests, such as the Sitka spruce forests in the Pacific North-west of North America (Winchester 1997). Using various trapping methods, abundance and species richness of arthropods were sampled in forest canopies, forest floors, transition zones and in clearcut areas, with emphasis on the beetle family Staphylinidae.

Although the greatest number of species was found in the disturbed transition zone, the fact that 62% of these were linked to forest interior species indicates the importance of this zone as a source of species. Principal components analysis confirmed the separation of the various zones and also confirmed the unique nature of the old-growth forest floor staphylinid fauna. In addition, canopy-dwelling oribatid mites were highly specific to old-growth canopies and it was concluded that following clearcut second-growth canopies were unlikely to provide the specialized habitat required for these species within the rotation period (up to 120 years) (Winchester & Ring 1996).

Another important habitat in old-growth forests is dead wood, which provides a specialized resource for saproxylic invertebrates. Comparison of the saproxylic Coleoptera of Scots pine and Norway spruce in old-growth and managed forests in central Finland provided some interesting contrasts between species numbers and rarity value (Vaisanen *et al.* 1993). Although numbers of both species and individuals were higher in managed forests (78 species and 1302 individuals compared with 55 species and 512 individuals), this was due mainly to the presence of bark beetles (Scolytidae). However, when species classified as rare were considered, the proportion in the samples was only 1% in managed forests compared with 12% in the old-growth forests. The authors concluded that dead-wood faunas in managed forests, although providing high biodiversity value, were virtually completely different from those of old-growth forests. This was almost entirely at the expense of relatively rare old-growth dead-wood specialists that are not likely to colonize managed forests, particularly if the degree of fragmentation of old-growth forests is large. Such findings are repeated in other studies of old-growth and managed forests, confirming that it is not just the absolute amount of dead wood that is important but also the resource pool in the surrounding area, especially for specialist saproxylic species with limited dispersal potential (Speight 1989; Albrecht 1991).

The processes that determine the interactions between forests and their associated invertebrate faunas are therefore complex and it is important that the implications of forest management practices, for whatever purpose, are taken into account in assessing potential impacts on invertebrate species richness and abundance. Some of the management options that can be considered at a range of scales have been alluded to in this chapter. It is not always apparent whether forest pest management in practice has taken account of the lessons that can be learned from existing ecological knowledge and principles. However, the fact that tools such as GIS, GPS and portable computers are now available should provide the impetus to develop more context-sensitive and realistic decision support systems for future management of forest invertebrates.

# 7

# Forest Soils

## INTRODUCTION

To write a chapter on forest soils in detail at a global level is an impossible task. To give the general reader an appreciation of the scope of the subject and indeed to acknowledge its historical roots, we list some major publications that have dominated the field. In the USA, an enormously successful book on soil science has been *The Nature and Properties of Soils*, a text that has evolved with the science over 11 editions (Brady 1996), starting from the earliest by T.L. Lyon and H.O. Buckman in the 1920s. Similarly in the UK, EJ. Russell's *Soil Conditions and Plant Growth*, first published as a series of monographs in 1912, is also in its 11th edition (Wild 1988). Both are concerned essentially with soils and agriculture. We also cite Jenny's (1941, 1980) pioneering works on soil formation and soil behaviour that have been of enormous influence in ecological research, including conceptual models of the rate of accumulation of soil carbon.

Texts on forest soils have been smaller and more regional. For example, in the USA there is a distinguished series of texts by Lutz and Chandler (1946), Wilde (1946), Pritchett (1979) and Binkley (1986); in Canada by Armson (1977); in Russia by Remezov and Pogrebnyak (1969); in Australia by Attiwill and Leeper (1987); to name a selective few. Again, if we were asked to name but a handful of publications in addition to texts such as these that have been seminal in the historical development of the modern science of forest soils and nutrient cycling we would include the works of Ebermayer (1876), Rennie (1955) and Likens *et al.* (1977).

## Scope of this chapter

If we were asked 'In what ways do forest soils differ from other soils?', our first response would be that soil processes in forest soils are dominated by the biological search for C, the result of continued input to the soil of C from litterfall and root turnover. We have therefore chosen this as a dominant theme in discussing soil processes. We then address some specifically forest-related ecological and management issues. However, in both of these attempts we can only be selective rather than comprehensive in our treatment, and no doubt our Anglo-Saxon origins and our scientific biases are all too apparent.

From the 1950s through the 1980s, most of the development in forest soils centred around soil chemistry and nutrient cycling. Throughout much of that era, the concentration was on above-ground pools and processes. More recently, this imbalance has begun to be redressed and the contribution of fine-root turnover to nutrient cycling and of the rhizosphere to nutrient uptake are now major areas of research in forest soils. Nutrient cycling, roots and root turnover, and the rhizosphere are the first themes of this chapter.

There is probably general agreement that the two elements of most immediate concern are N and P, with N receiving greater attention, particularly in the Northern Hemisphere. Mineralization is the process by which the nutrients 'trapped' in organic matter, cycled from plant to soil, are made mineral again and available as simple ion for uptake. Mineralization is the gateway that controls the rate of nutrient cycling.

Thus mineralization, litter decomposition and soil organic matter are inextricably linked. Where the oxidation of C compounds and the availability of nutrients are 'driven by nature' (Cadisch & Giller 1997), these linkages are fundamental to our understanding of the rate of nutrient cycling, of the forms, storage and turnover of C in forest soils, and of the response of soil C to changing environments. These are the secondary themes of this chapter.

A great deal of research in forest soils has focused on the effects on forests of increasing acidic inputs from highly industrialized societies. Acidic inputs have been associated with both forest decline (mainly through S deposition) and forest enrichment (through N deposition). Forest decline and 'N-saturation' are the third themes of the chapter.

Our last themes address the conservative and sustainable management of our forests for timber. Do our silvicultural and logging systems result in a loss of soil fertility and is that loss significant for sustained productivity? Do these systems compact the soil and is compaction significant for sustained productivity?

Our final limitation in setting the scope of this chapter is that forests grow on all soil types from the deepest sands to the heaviest clays. Soils are classified in numerous ways, using various features, at various scales, for various purposes. Soil classification is therefore too large a topic for us to cover. For an international audience, the type of soil is generally described by either the USA Department of Agriculture comprehensive classification (Soil Survey Staff 1996) or the World Soil Classification (FAO-UNESCO 1988).

## NUTRIENT CYCLING IN FORESTS

### Soil chemistry and nutrient cycling

By and large, the study of soil chemistry has concentrated on agricultural soils. The immediate practical questions to be answered have centred on the availability of plant nutrients, the need to develop chemical tests to define those soils suitable for particular crops and the prescription of the quantity of fertilizer that must be added to increase the productivity of the crop. With the exclusion of N (the availability of which is driven by biological processes in all soils), we might summarize the essential difference between the availability of plant nutrients in agricultural and forest soils:

- the availability of nutrients in agricultural soils is an immediate concern that mostly addresses inorganic equilibria over weeks or months;
- the availability of nutrients in forest soils is a long-term concern that should address biological processes and inorganic equilibria over years and centuries.

The growth of forests is a long-term process in which nutrients are cycled from plant to soil in litterfall and root turnover. Nutrients are withdrawn from tissues as they age and are translocated to actively growing tissues. Timber harvesting takes away the aged tissues with relatively low nutrient concentrations, leaving the tissues with relatively high nutrient concentrations on site. In contrast, all of the nutrient supply for an annual crop is taken up from the soil within the few months the crop takes to reach maturity.

There is little cycling of nutrients from plant to soil. At maturity, the most nutrient-rich parts of the crop (seeds, leaves, storage organs) are harvested, leaving only the tissues with relatively low nutrient concentrations behind.

Thus, from the pioneering work of Ebermayer (1876) more than a century ago, much of the work on forest soils has been directed towards nutrient cycling, in particular:

- quantifying the cycle of nutrients (uptake and return between plant and soil and retranslocation within the plant);
- quantifying inputs and outputs, including nutrient removals in harvested timber.

While all of this is obvious, the science required is not obvious. Twenty years ago Stone (1979) concluded that despite a wealth of data on nutrient cycling processes for many forests, we do not have a sound experimental basis on which to assess the sustained productivity of forest ecosystems. Ten years later, Landsberg *et al.* (1991) restated that conclusion: 'Conventional chemical analyses to determine the nutritional status of the soil yield information that may have little relevance in calculating the capacity of the soil to provide nutrients to trees, and the complexity of uptake processes through complex, dynamic and poorly defined root systems is tremendous.'

Rather than attempting to review ongoing research on the experimental and analytical basis for assessing sustainability of nutrient supply, we give only a factual account of what is known of the chemistry and cycling of the major nutrients. We do this in general terms, using data for a temperate forest of above-average productivity and we include a summary of the form, function, concentration and cycling of the major nutrients N, P, Ca, Mg and K.

*Biomass*

The above-ground mass of the forest (trees, under-storey and shrub layer) may reach 500t ha$^{-1}$. Litterfall *(A)* is 8tha$^{-1}$ year$^{-1}$ but may reach 10tha$^{-1}$ year$^{-1}$. If the litter layers (*Q*) weigh 30tha$^{-1}$, the decomposition rate of litter $k = A/(A + Q) = 0.21$ year$^{-1}$, and the half-life $t_{0\ 5} = 0.693/ft = 3.3$ years. This and subsequent analyses are based only on the above-ground stand. Data for root mass and root turnover are both more limited and variable, and have been intensively summarized (Vogt *et al.* 1986, 1996). We have previously estimated a rate of root turnover for our temperate forest of above-average productivity of 2-3tha$^{-1}$ year$^{-1}$ (Attiwill & Adams 1993).

*Nitrogen*

The concentration of N in leaves is greater than the concentration of any other nutrient, but much of this N is withdrawn from leaves before they die and fall as litter. Internal redistribution of N accounts for almost 30% of the total annual above-ground cycle. The half-life of N during litter decomposition is 3.5 years, a little longer than that for dry weight, so that N is immobilized during decomposition of the litter. The input of N in rainfall of 5 kg ha$^{-1}$ year$^{-1}$ is more or less balanced by the output of N in streamwater. More than 95% or so of N in the soil is in organic form; organic N is mineralized to $NH_4^+$-N and then under certain conditions to $NO_3$—N, a form freely available for uptake by plants. $NH_4^+$-N is held on negatively charged exchange sites of soils, but $NO_3$—N is mobile within the soil and is the form in which N moves with drainage waters into streams.

In soils of natural forests of the world, the generally high C/N ratio of litter and soil organic matter precludes mineralization proceeding to $NO_3$—N, so that the dominant form of N is $NH_4^+$-N (Jansson 1958; Wollum & Davey 1975; Raven *et al.* 1978; Attiwill *et al.* 1996). The cycling of N is therefore conservative, and N may be immobilized in litter of older forests (particularly in cooler-temperate areas) to the extent that the trees become N-deficient (Tamm 1964). Disturbance (treefall and the creation of gaps, fire, harvesting for timber) provides the conditions that result in a decrease in C/N ratio and the production of $NO_3$—N, in which form N may be leached from the soil into streamwater. Rapid regeneration of the forest, including rapid uptake of $NO_3$—N, following disturbance is therefore the key to the conservation of N (Weston & Attiwill 1990, 1996).

*Phosphorus*

Phosphorus occurs in highly mobile forms within plants and, like N, is withdrawn from plant parts during senescence; internal redistribution of P accounts for 50% of the total annual above-ground cycle. The half-life of P during decomposition is 4.1 years, i.e. P is immobilized throughout decomposition to a greater degree than N. A negligible amount of P comes in with the rain. P as $H_2PO_4^-$ is adsorbed by clay colloids and is then highly immobile in the soils so that losses of P to streamwater are negligible (see Table 7.3). However, in forest soils more than 50% of the total P in surface soils may be in organic form. Mineralization of organic P is therefore fundamental to ecosystem function, and P availability is 'determined by competition between biological and geochemical sinks for phosphate anions' (Attiwill & Adams 1993). Whereas methods for estimating rates of N mineralization have been developed and are in general application (e.g. Raison *et al.* 1987; Adams *et al.* 1989), there are no methods for routinely estimating the rate of P mineralization. The difficulty is that whenever $H_2PO_4^-$ comes into solution, there is a surface to adsorb it. This competitiveness between biological and chemical sinks for P remains a priority area of research in quantifying the long-term sus-tainability of P supply for forests.

*Calcium, magnesium and potassium*

Ca, Mg and K, and the non-essential element Na, are generally grouped as the major exchangeable cations in soil. An exchangeable cation is held adsorbed on a negatively charged surface of a colloid (both inorganic and organic). In this exchangeable form, the cations are not soluble but are brought quickly into solution and made available for uptake by exchange of protons from plant roots.

Ca is highly immobile within the plant, where it is incorporated within cell walls. Thus 80% of the total annual above-ground cycle of Ca is in lit-terfall. In contrast, K is highly mobile and $K^+$ easily leached from tree crowns by rain; leaching accounts for about 30% of the total annual above-ground cycle of K.

### *Trace elements (micronutrients)*

The trace elements Fe, Mn, Cu, Zn, Mo and B have all played spectacular parts in agriculture and in tree plantations where plants have been introduced into new land. Applications of trace amounts (e.g. as little as 0.5kgha$^{-1}$ of Mo) have corrected deficiencies that would otherwise have made agriculture or plantation forestry unproductive. We know of no such experience in natural forests, although there is little doubt that the distribution of many species may be determined by their ability to increase or restrict the supply of one or more of the trace elements. For example, concentrations of Mn in leaves and bark differ consistently and significantly between groups of the genus *Eucalyptus* (Lambert 1981; Judd *et al.* 1996) but whether these differences are of significance in determining the distribution of species is unknown.

## Roots and the rhizosphere

Roots supply plants with water and nutrients and provide anchorage in soil. Most studies of roots in forests have been directed to the surface horizons, where fine roots (< 2 mm in diameter) are concentrated and where water and the bulk of nutrients are sourced. There are few studies of roots below 1 m, with only nine reported in the review of Jackson *et al.* (1996). In general our knowledge of root characteristics, including biomass, distribution and fine-root area, is poorly developed compared with the above-ground components of forests. Much of the recent interest in root distribution and turnover in forests has arisen primarily from concern for increased C flux from soils due to climate change (Jackson *et al.* 1996, 1997; Silver 1998).

### *Distribution, mass and turnover of roots*

A global analysis of root distribution in terrestrial biomes (Jackson *et al.* 1996) showed that boreal forests have the shallowest rooting profiles (80-90% of roots in the upper 30 cm of forest soils) and that temperate coniferous forests have the deepest rooting profiles (52% of roots in the upper 30 cm). When considered together, all temperate and tropical trees have 26% of roots in the top 10 cm and 60% in the top 30 cm (Jackson *et al.* 1996). Root biomass of forests has been estimated at 2-5 kg $m^{-2}$, with the highest values in tropical evergreen forests (Jackson *et al.* 1996). However, discrepancies between model estimates and measured characteristics of fine-root length and surface area point to the need for more data on root biomass and distribution, and for more careful measurement of fine root biomass in particular.

These analyses of root distributions and characteristics (biomass, surface area and nutrient content) among forests will enable improved modelling of water and nutrient uptake on a global scale (Jackson *et al.* 1997). They also provide the basis for improving global models of the impact of climate change on vegetation distribution and C sequestration in soil. When linked with models that predict the change in vegetation distribution and type, they will allow better prediction of the consequences of global environmental change (Jackson *et al.* 1996). The need for this type of analysis seems all the more urgent when it is considered that roots can account for more than 50% of net primary productivity (NPP) in forests (Vogt *et al.* 1986; Raich & Nadelhoffer 1989; Jackson *et al.* 1997) and that the global pool of C in fine roots is 5% of the size of the atmospheric C pool. Because about half the global pool of C in fine roots is in forests, further study of factors influencing fine-root turnover is critical to the understanding of $CO_2$ flux between terrestrial and atmospheric C pools.

While roots are concentrated in the upper metre of soil in most forests, deep roots become important in forests subject to periodic drought because they provide access to water held deep in the soil (Stone & Kalisz 1991; Nepstad *et al.* 1994). Carbon from these deep roots may contribute over decades to net C release from soil following conversion of forest to pasture, as shown by Nepstad *et al.* (1994) in an Amazonian rain forest. These sources of deep soil C have not been generally considered in estimates of C flux from forest soils, especially those of tropical regions where forests continue to be cleared for agriculture.

The turnover of fine roots has been investigated in relation to forest disturbance, including clearcutting, in the Northern Hardwood ecosystem (Fahey & Hughes 1994). Biomass of fine roots (<2mm) in mature hardwood forest was 471 $gm^{-2}$, with average lifespan ranging from 8 to 10 months. Biomass of fine roots recovered to 71% of that in the mature forest within 3 years after clearcutting (Fahey & Hughes 1994).

### *Rhizosphere*

The rhizosphere is the area of soil immediately surrounding the root where biotic and abiotic processes interact to create an environment distinct from bulk soil further from the root. Compared with bulk soil, rhizosphere soil is usually more intensively weathered, has a lower pH and has greater concentrations of cations and P (Griffiths *et al.* 1994; Gobran *et al.* 1998). Exudates released from plant roots and microorganisms benefit plant growth by increasing nutrient acquisition and metal detoxification, by alleviation of anaerobic stress in roots and by their action in mineral weathering (Illmer *et al.* 1995; Jones 1998).

There is no doubt that some forest species can modify the rhizosphere to access previously unavailable reserves of soil P by releasing organic acids that solubilize inorganic P. For example, Grierson (1992) demonstrated that in P-deficient soils, *Banksia integrifolia* induced the development of short-branched, tertiary, lateral roots (proteoid or cluster roots) that released citric and malic acids (the main organic acids induced under P deficiency). Organic acids probably increase the availability of P in soils through a combination of decreased adsorption of P, increased solubilization of P compounds and chelation of metals such as Fe (Bolan *et al.* 1994). Other rhizosphere processes that impact on the availability of soil P reserves include increase in root hair length and density, release of C that enhances mycorrhizal exploitation of bulk soil, and the release of phosphatases that solubilize organically bound P (Jones 1998).

The role and function of mycorrhizas in forest soils has been reviewed by Brundrett (1991) and Vogt *et al.* (1991) and there is no doubt that mycorrhizal associations benefit trees by enhancing nutrient acquisition, especially in relation to P. The evidence for N and P transfer to forest trees via mycorrhizal associations has been demonstrated in tracer studies run over relatively short time-courses. However, it is only recently that new techniques have allowed longer-term studies; these have confirmed the benefits of mycorrhizas to nutrient acquisition and have also demonstrated increased plant growth as a result (Brandes *et al.* 1998). Further work of this kind is required to establish the relative contribution of mycorrhizal uptake mechanisms to nutrient acquisition in mature forests over the long term.

The activity of microorganisms in the rhizos-phere is driven by the release of low-molecular-weight organic compounds by the root and there is some evidence to suggest that the growth of certain microorganisms is favoured while that of others is suppressed. For example, Gonzalez *et al.*(1995) reported that while microbial numbers were greater in the rhizosphere of *Alnus*, roots differentially favoured colonization by proteolytic and ammonifying organisms and inhibited nitrifying organisms. Norton and Firestone (1996)reported a greater than 50% increase in N turnover (mineralization and immobilization) in the rhizosphere relative to bulk soils in *Pinus ponderosa* microcosms. Further evidence of the complex interactions in the rhizosphere come from reports of ectomycorrhizal fungi stimulating bacterial growth associated with the release of citric acid by the fungus and, conversely, of inhi bition of bacterial growth (Olsson & Wallander 1998).

Recent work has aimed at characterizing the chemical and biotic nature of the rhizospere, especially as it is influenced by soil liming (Clemensson-Lindell & Persson 1993; Hüttl & Schneider 1998) and nutrient addition (Majdi & Bergholm 1995; Teng & Timmer 1995). Most work has been focused in the vicinity of industrialized areas of the Northern Hemisphere, where there are concerns for the unimpeded functioning of beneficial root-rhizosphere reactions under both controlled (fertilizer inputs) and uncontrolled (atmospheric deposition) inputs of acidity, N and S (Gorissen *et al.* 1994; Majdi & Rosengren-brinck 1994). Reported effects of ammonium sulphate application on rhizosphere chemistry vary and many of the changes, such as increased acidity and Al (Majdi & Rosengrenbrinck 1994; Rosengrenbrinck *et al.* 1995), are found in bulk soils as well (Majdi & Persson 1995). Other effects include depletion of P and K from the rhizosphere of Norway spruce in southern Sweden induced by ammonium deposition and the subsequent stimulation of P and K uptake (Clegg & Gobran 1997).

The benefits of rhizosphere organisms, including mycorrhizal fungi, in maintaining the growth of tree seedlings in nurseries has long been known. This concept has been extended recently to the development of soil inoculation techniques to improve the growth of seedlings in the reforestation of degraded or clearcut sites (Chanway 1997), where inoculation with forest soil has increased the number of ectomycorrhizas (Colinas *et al.* 1994a). Differential effects on survival and growth of seedlings, where planting holes were inoculated with soil transferred from forestry or plantation sites, have been reported (Colinas *et al.* 1994b). Increases in seedling growth have been associated with increased nutrient mineralization brought about by soil animals stimulating microbial turnover in transferred soil, while increases in survival probably resulted from the introduction of beneficial rhizosphere organisms (Colinas *et al.* 1994b).

Another dimension in the use of mycorrhizas to benefit tree growth has been the recognition by Garbaye (1994) of bacteria that selectively promote mycorrhizal development (the so-called 'mycorrhization helper' bacteria, MHB). In anticipation of the applications discussed by Chanway (1997) the process of selecting and using MHB cultures for controlled 'mycorrhization' has been patented by Garbaye and Duponnis (1991).

## LITTER AND SOIL ORGANIC MATTER

### Litter and litter decomposition

There have been several syntheses of rates of accession and decomposition of litter in forests over the last two decades (Schlesinger 1977; Vogt *et al.* 1986; Mathews 1997). Large amounts of litter fall from the above-ground parts of forests to the soil annually; this litter is a major determinant of nutrient cycles in forests (Attiwill & Adams 1993). On a global scale, annual above-ground litter production in forests, woodland and wooded grassland is estimated at about 39 Pg of dry matter (1Pg= $10^{15}$g), which is a little less than 50% of total global litter production (Mathews 1997). Annual above-ground litter accession is highest in tropical, broadleaved, evergreen forests at 15300kgha$^{-1}$ and as low as 130 kg ha$^{-1}$ in boreal, needle-leaved, evergreen forests (Vogt *et al.* 1986).

On a global scale, average litter turnover time is about 5 years for litter and about 13 years for coarse woody debris (usually classified as materials >7cm diameter) (Mathews 1997). As decomposition of forest litter is an important component of global C cycling, the potential impact of climate change on C release from litter has attracted considerable attention over the last decade (McHale *et al.* 1998; Moore *et al.* 1999). In a recent review on a global scale of litter decomposition (measured as decomposition in the first year), Aerts (1997) concluded that climate (expressed as actual evapotranspiration) is a better predictor of the rate of litter decomposition than litter chemistry, although climate indirectly influences litter chemistry. However, litter chemistry became the best predictor of decomposition rates within discrete climatic regions, especially in the tropics and in the Mediterranean region (see also Gallardo & Merino 1999) where the ratio of concentration of lignin to concentration of total N was the best indicator of litter decomposition (Aerts 1997). The strong correlation between decomposition rates and the lignin :N ratio of litter has been corroborated in a wide range of Canadian forests (Moore *et al.* 1999) where 73% of the variance in decomposition was accounted for by a multiple regression of mean annual temperature, mean annual precipitation and the lignin: N ratio of the litter.

On a regional or global scale, the rate of litter decomposition in forests is predominantly controlled by climate (temperature and moisture, as shown in general terms by a strong negative correlation with latitude) and by the quality of litter in terms of how beneficial it is to the micro-bial community as a source of energy or nutrients (Meentemeyer 1978; Meentemeyer *et al.* 1982; Stohlgren 1988; Currie & Aber 1997). High concentrations of nutrients relative to stored energy promote more rapid decomposition rates, so that C: N and lignin: N concentration ratios are commonly used as variables for litter quality (Harmon *et al.* 1990; Parton *et al.* 1994). Across the generally mesic forested biomes, climate and litter quality seem to be the best indicators of the rate of litter decomposition. However, the value of litter quality as a determinant of the rate of litter decomposition does not hold across all biomes, and is poorly correlated with the rate of decomposition in desert regions (Schaefer *et al.* 1985).

While accession of exogenous N to forested land has raised concern for increased decomposition of litter, reports of effects on decomposition rates vary. For example, Kuperman (1999) reported increased rates of litter decomposition (and of N mineralization) in white oak stands of the lower midwestern USA following increased N deposition. However, in a study encompassing a wide range of sites, Prescott (1995) found that increased N availability following fertilizer addition or deposition did not alter rates of litter decomposition in forests.

It seems clear that while exogenous N may increase decomposition in some forests, especially over the first year following litterfall, it may also increase the lignin content of litter and reduce long-term decomposition rates in other forests (Magill & Aber 1998).

The reduced decomposition of N-rich litter may be due to the chemical formation of stable nitrogenous compounds from lignin by-products (Berg *et al.* 1995). These N compounds are thought to form by chemical condensation reactions and may be highly resistant to biological degradation (Aerts 1997). Current evidence therefore suggests that an increased rate of N deposition decreases the rate of C release in the latter stages of decomposition and stimulates C storage, both in increased woody biomass and through soil humus formation and, thereby, sequestration of C as soil organic matter (Berg 1986; Magill & Aber 1998).

Below-ground inputs of C and nutrients in forests may become the dominant pathway of nutrient return, especially in cold-temperate climates (Vogt *et al.* 1986). Berg *et al.* (1998) studied root litter decomposition in coniferous forests from northern Scandinavia to north-eastern Germany and found little correlation between the loss of mass in the first year and climatic variables when different forest types were considered together. When the *Pinus* and *Picea* stands were treated separately, stronger correlations emerged with initial concentration of P and July temperatures explaining 71% of the variation in the rate of decomposition of root litter of *Picea* stands along the transect.

Soil-warming experiments in a northern hardwood forest, where the forest floor was heated with a cable, showed that the $CO_2$ flux from leaf litter increased exponentially with increasing soil temperature (McHale *et al.* 1998). These increased rates of $CO_2$ flux from litter decomposition may represent a short-term depletion of labile C with increasing temperature, so that the response diminishes over time (Peterjohn *et al.* 1994). However, if global warming causes increases in the rate of biomass production and in the rate of litterfall, the rate of litter decomposition will increase and increases in the rate of $CO_2$ emissions from forest litter will be sustained (McHale *et al.* 1998).

## Carbon in forest soils

### *Forest soils and carbon storage*

The global cycling of C involves fluxes between the fossil C reservoir, the atmosphere, the oceans and the terrestrial biosphere (Schimel 1995). The role of the terrestrial biosphere in buffering or contributing to increases in $CO_2$ of the global C cycle is uncertain, partly because of the difficulty in predicting the likely impact of climate change, associated with rising atmospheric $CO_2$, on C flux (Smith *et al.* 1993). Because forests and wooded lands cover about one-third (4100 million ha) of the land area of the earth, they are a significant component of that C stored in the terrestrial biosphere which is exchangeable with the atmospheric pool of C (Sedjo 1992; Dixon 1994; Dixon *et al.* 1994). The estimated 787 Pg of C in forest soils and associated peats is more than two-thirds of the total of 1146Pg of C held in forests (Dixon 1994; Dixon *et al.* 1994).

About 33% of forest soils are covered by coniferous forests in the high latitudes of Russia, Canada and Alaska. About 25% of forest soils are covered by forests in the mid-latitudes of continental USA, Europe and Australia, while the low-latitude (tropical) forests of Asia, Africa and the Americas account for 42% of forested land. More than half of the low-latitude forests are in tropical America. The amount of C stored in forest soils increases with northern latitude, so that low-latitude (tropical) forests contain about 27% of the pool of soil C, increasing to about 60% in soils of high-latitude northern boreal forests (Dixon 1994; Dixon *et al.* 1994). Thus the soils of high-latitude northern forests are an important

C pool and any change in the climate or the management of these forests that diminishes the soil C pool will significantly affect global C storage (Dixon 1994; Dixon *et al.* 1994; Bird *et al.* 1996). This applies equally to the tropical forests of lower latitudes, where soil C stocks are less than half those of the boreal forests, but where small increases in temperature will have a large proportional effect on metabolic activity (McKane *et al.* 1995).

The balance between net production or consumption of $CO_2$ in forests is largely determined by decomposition rate and the rate of soil organic C formation. This balance can be estimated reasonably well for undisturbed forests (Van Breemen & Feijtel 1990). However, following clearing of forests the relatively low-level sinks of $CO_2$ in forest soils become high-level sources. Conversely, reforestation may restore the sink capacity of the soil, thereby off-setting to some extent the increase in atmospheric $CO_2$ due to land clearing and the oxidation of fossil fuels. Extensive tree planting and the development of policies for 'carbon credits' has therefore focused research efforts on the amount and stability of soil C and on processes that control the inputs and outputs of C.

Because soils contain the largest pool of C in terrestrial ecosystems, small changes in soil temperature or moisture or both could lead to a substantial increase in atmospheric $CO_2$ concentrations (Anderson 1992). Current general circulation models predict that the temperature of the earth's atmosphere will increase by 1°C over the next 35 years (Houghton *et al.* 1992), altering global climate patterns and hence impacting on many ecosystem processes. Climate warming is predicted to be greatest nearest the poles so that boreal and tundra biomes will be influenced the most; this has generated considerable interest in the interaction of climate with the biogeochem-istry of these biomes (Rapalee *et al.* 1998).

Carbon accumulates in forest soil as a result of litterfall and root input, while outputs result from microbial degradation of organic matter, eluvia-tion, solution losses and erosion. The carrying capacity for soil C increases as forests mature, and the maximum carrying capacity is controlled by climate, topography, soil type and vegetation (Dewar 1991; Van Cleve & Powers 1995).

For example, Niklinska *et al.* (1999) measured the effect of increasing temperature on the rate of respiration of humus from forests of Scots pine and predicted an increase in the annual rate of soil respiration for the high latitudes of the Northern Hemisphere of 70 Tg (1Tg=$10^{12}$g) $CO_2$ with a 2°C increase in temperature. Niklinska *et al.* (1999) therefore suggested that northern forests may become a net C source with increased global temperatures.

Much of the global store of soil C is sequestered in peats and peaty soils developed in cold and wet climates (Harrison *et al.* 1995; Hart & Perry 1999). Drying of these organic soils as a result of climate change will significantly increase the rate of loss of soil C (Schimel *et al.* 1994), although the magnitude of change of C-flux following the drainage of peatlands and their conversion to conifer plantations has been difficult to measure (Cannell *et al.* 1993).

In general, the state of current knowledge is insufficient to predict whether rising temperatures will lead to a net release or storage of C in the terrestrial biosphere (Post & Pastor 1996; Houghton *et al.* 1998). The potential for forest soils to store or release C under changing climate and land use has been modelled on a large scale by Schimel *et al.* (1994). The CENTURY model of the turnover of soil organic matter applied to global forest and grassland ecosystems showed that soil texture and lignin concentration of foliage are key variables influencing soil C storage. Feedback between the N cycle and NPP influences model predictions of the terrestrial C balance (Schimel *et al.* 1994). All projections, based on a range of ecosystem models, suggest that increasing $CO_2$ will increase C inputs to soil, which will in turn widen the detrital C: N ratio, thereby decreasing the rate of decomposition. Simulations using the CENTURY model show that N cycles more rapidly under warmer temperatures and that as soil organic matter is lost more N becomes available for plant growth, resulting in more soil organic matter formation, and thereby providing negative feedback (Schimel *et al.* 1994). It seems that the efflux of organic C from soil with global warming would be much greater, perhaps double the current efflux, without this so-called 'negative feedback' (Schimel *et al.* 1990, 1994). However, speculation on the role of increased N mineralization from soil organic matter in increasing the vegetation biomass has not been substantiated with field data (Houghton *et al.* 1998).

The Intergovernmental Panel Assessment on Climate Change (IPCC) assessed the role of forests in the terrestrial C cycle over the decade 1980-90. The assessment concluded that regrowth forests of the Northern Hemisphere may have been a net sink of 0.5Gtyear$^{-1}$ of C, while deforestation in the tropics released 1.6Gtyear$^{-1}$ of C (Harrison *et al.* 1995). Climate change in tropical regions also has the potential to alter C storage significantly, as shown by Tian *et al.* (1998) in a study of the effect of interannual climate variability on C storage in Amazonian ecosystems. The hot dry weather of El Niño years in Amazonia decreased soil moisture, leading to a net release of 0.2Pg of C in 1987 and 1992. In other years, the forests and savannahs of the Amazon basin are a sink of up to 0.7Pg annually of C (Tian *et al.* 1998). These findings are supported by Silver (1998) who predicted that small increases in temperature and $CO_2$ will result in more rapid decomposition and therefore an increased $CO_2$ flux from moist tropical soils.

Deep soils of tropical forests contain a significant proportion of the ecosystem C, with estimates of the mass of C at depths below 1 m equal to or greater than the mass of C in the above-ground forest (Nepstad *et al.* 1994). This C, held deep in the soil, represents a significant source of $CO_2$ emissions following deforestation and conversion to pasture.

The impacts of forest regrowth, of increasing atmospheric concentrations of $CO_2$, of N deposition and of global warming on C storage in forests remain critical areas for future research. Holland *et al.* (1997) modelled the effects of atmospheric $NH_X$-N and NO -N deposition on global terrestrial C and estimated an increased rate of storage of C of 1.5-2.0Gtyear$^{-1}$, an amount similar to that of the missing terrestrial sink. Makipaa (1995) found increases in C storage in both humus (14-87%) and mineral soil (15-167%) following long-term N addition experiments in boreal forest. In temperate forests, N deposition is expected to both alleviate N deficiency and increase C sequestration in biomass and soil organic matter (Nilsson 1995).

Houghton (1996) claimed that the predicted increase in mid-latitude forest sequestration of C has occurred (data from Canada,

USA, Europe and the former USSR), with a greater than expected accumulation of C (0.8Pgyear$^{-1}$) in biomass and soil pools during the 1980s. These gains of C in forests of the mid-latitudes have been offset by deforestation in the tropics, which contributed 1.6Pgyear$^{-1}$ of C during the 1980s (Dixon *et al.* 1994; Houghton & Hackler 1995; Schimel *et al.* 1995; Houghton 1996).

Forest management to conserve and sequester C has been identified as one of three main strategies to stabilize greenhouse gases (Dixon *et al.* 1996). Johnson *et al.* (1995) likewise identified significant global-scale opportunities for both reducing C emissions from soil and increasing C sequestration in soil. The viability of such strategies has already been questioned on both sociological and scientific grounds (Binkley *et al.* 1997). Management to increase C storage in forests has been investigated at regional, national and global levels (e.g. Dixon 1995; Carter *et al.* 1998; Dick *et al.* 1998; Liski *et al.* 1998). In a global survey of 94 nations, Dixon *et al.* (1994) found that the establishment of agroforestry and alternative land-use systems (fuelwood and fibre plantations, bioreserves, intercropping systems, shelterbelts and windbreaks) on marginal and degraded lands could sequester 0.82-2.2Pg annually. The estimated cost of these activities was $US1–69 per tonne of C, which compared favourably with other options.

Deforestation, especially of low-latitude forests, remains one of the main causes of the global net release of $CO_2$ from forest soils to the atmosphere, yet the magnitude of emissions is uncertain due to limited studies of changes in soil C following forest clearing (Veldkamp 1994). Dixon (1994) and Dixon *et al.* (1994) estimated that 15.4 x $10^9$ha of forest were lost in the two decades between 1971 and 1990; the release of C from these soils will continue for decades as the C content of the soil gradually stabilizes following the change in land use (Veldkamp 1994).

Torn *et al.* (1997) correlated the content of soil organic C with soil age in a forested chronose-quence in Hawaii and concluded that the accumulation and subsequent loss of organic matter was largely driven by changes in the cycling of mineral-stabilized C over millennia rather than by changes in either the amount of fast-cycling organic matter or NPP Soil mineralogy is therefore important in determining the quantity of organic C stored in soil, its turnover time, and C exchange between ecosystems and the atmosphere during long-term soil development. The C storage capacity of the 25% of the world's mineral soils that began developing after the last major glaciation will change as they undergo further development, and they may eventually become long-term sources of atmospheric $CO_2$ (Torn *et al.* 1997).

*Forms and turnover of carbon in soil*

Organic matter includes most of the litter and humic layer in forest soils and generally forms 1-10% of surface mineral horizons (Attiwill & Leeper 1987). The comminution and decomposition of above- and below-ground litter in forests releases $CO_2$, and nutrient elements are mineralized and immobilized. While all organic matter is ultimately oxidized in forest soils, compounds that decompose over centuries and millennia are synthesized in the process of humification, which involves processes of both degradation and synthesis. This humus is amorphous and is intimately mixed with soil mineral components; it is difficult to describe in any precise chemical manner. The inert organic matter of forest soils is protected from decomposition in a number of ways, including association with mineral particles, which is the basis for longer turnover times for organic matter in clay-textured soils relative to sandy soils. Most of the organic matter in soils is very old, and decomposition to inorganic constituents will take centuries to millennia (Jenkinson & Rayner 1977; Attiwill & Leeper 1987; Schimel *et al.* 1994).

Both chemical and physical methods have been used to fractionate soil organic matter; however, it has not been possible to separate organic matter into discrete components of varying decomposability (Sanchez *et al.* 1989). Traditional chemical methods date from the classical empirical fractionation based on sequential acid and alkali extractions; this recognizes fulvic acids (soluble in both acid and alkali), humic acids (soluble in alkali and precipitated by acid) and humin (insoluble in both alkali and acid). The structure of humic compounds is not accurately known but humus formation involves the polymerization of aromatic compounds, many of which are derived from lignin and polyphenolic pigments. The application of analytical techniques such as nuclear magnetic resonance, Fourier transform infrared spectroscopy and pyrolysis-mass spectroscopy are refining our understanding of the structure and reactiveness of humic compounds in soil (Schnitzer & Schulten 1995). Recent studies of humic compounds in forest soils have shown that the stability of the inert or occluded organic matter appears to be inversely related to the content of *O*-alkyl C and directly related to the content of aromatic C (Golchin *et al.* 1995).

Physical methods based on density usually separate soil according to free particulate organic matter (macro-organic matter), occluded particulate organic matter (light fraction) and colloidal or clay-associated organic matter (heavy or humified fraction) (e.g. Golchin *et al.* 1994). Density fractionation, based on dispersal of soil in NaI solution, has proved useful in separating organic matter into components that are physically and chemically distinct (Strickland & Sollins 1987), and which are mineralized at different rates by soil microorganisms (Christensen 1992; Boone 1994). Generally the light fraction extracted in NaI comprises partially decomposed roots and litter and is relatively labile; it has been reported to represent 25-50% of total soil carbon in forests of the United States (Strickland & Sollins 1987).

By contrast, the heavy fraction is regarded as more resistant humic material that is adsorbed on to mineral surfaces (Theodorou 1990). The heavy fraction can be further separated into proximate fractions such as soluble fats and oils, celluloses and lignin, and phenolic compounds. Studies employing these density-fractionation techniques (e.g. Entry & Emmingham 1998) have established that old-growth forests store more C than young-growth forests and that soils of old-growth forest contain greater percentages of recalcitrant organic matter typically richer in lignin and tannic compounds.

Further application of these methods, which identify the most biologically active forms of C in forest soils, are critical to developing a better understanding of soil organic matter turnover under a range of influences.

Turnover time ($1/k$) is defined as the time that C is held in organic forms per unit volume of forest floor and soil; it is therefore the time taken for all of the organic C in the soil to be replaced. Turnover times range from hours and days for freshly added and easily decomposed organic matter to centuries and millennia for highly stabilized organo-mineral complexes in the soil.

As apparent from the discussion of C storage in forest soils, much of the recent interest in forms of organic matter in forest soils and their turnover time is in relation to the impacts of climate change and forest management on pools and fluxes of C. Models of soil C turnover have been used to predict the effects of climate change and of different land uses on soil C fluxes (e.g. Kelly *et al.* 1997; Bolker *et al.* 1998; Peng & Apps 1998; Peng *et al.* 1998). While a number of models have been developed, the most popular are those based on the CENTURY model (Parton *et al.* 1988) and the ROTH-26.3 model (Coleman & Jenkinson 1996). These models have been used to simulate changes in organic C of arable and grassland soils with accuracy (e.g. Coleman & Jenkinson 1996).

However, studies in forest soils must encompass organic matter with a wide range of decomposition constants. Although the C models were originally developed for agricultural soils (Jenkinson & Rayner 1977; Oades 1988), they have recently been adapted to model the more complex environment of forest soils by inclusion of modules to accommodate the forest floor and litter layer (Parton *et al.* 1994; Vitousek *et al.* 1994). An adapted CENTURY model predicted the pattern or ranking of decomposition of surface litter in a matrix of *Metrosideros* sites on Mauna Loa but did not predict the magnitude of differences among sites (Vitousek *et al.* 1994). The refinement of the CENTURY model as it applies to forest soils will doubtless demand a better understanding of the reactivity of organic matter on the forest floor.

Studies of the proportion of the large store of soil C that exchanges with the atmosphere within centennial and shorter time-scales (compared with time-scales of millennia) are critical to our understanding of the global C cycle (Torn *et al.* 1997). At the scale of landscapes and over geological time, Torn *et al.* (1997) demonstrated that the turnover time for the large, passive pool of soil C is controlled primarily by soil mineralogy.

## FOREST SOILS AND ACIDIC INPUTS

Sulphur emissions have increased since the start of the Industrial Revolution to the extent that 75% of global emissions of S (total about 105Mtyear$^{-1}$) come from anthropogenic sources. While emissions from natural sources are about evenly distributed between Northern and Southern hemispheres, 90% of anthropogenic sulphur emissions come from sources in the Northern Hemisphere (about 70Mtyear$-^{1}$ in the Northern Hemisphere and 7.7Mtyear$^{-1}$ in the Southern Hemisphere.

## Forest decline

Forest decline in the Northern Hemisphere was increasingly observed during the 1970s and 1980s and was initially linked with the increasing acidity of rain, the latter caused by nitric and sulphuric acid formed from emissions from industry, cars, etc. However, there are other sources of acidity such as dry deposition (including both direct gaseous absorption and particulate deposition), and the rates of dry deposition of $SO_2$ and $NH_4^+$ may exceed rates of deposition in rainfall. There are also many causes of forest decline other than acid deposition.

There is little doubt that acid rain (used here in a generic sense) is a cause of some severe forest declines, particularly when close to emission sources. The most extensive studies have been in Germany (e.g. Ulrich 1983; Schulze 1989; Schulze & Ulrich 1991) and the results can be simply summarized.

With increasing acidity, cation exchange capacity is reduced (leading to the loss of $Ca^{2+}$ and $Mg^{2+}$) and the concentration of Al ions increases. Ca : Al and Mg : Al ratios in needles of *Abies alba* from healthy stands are greater than those in needles from stands that show symptoms of forest decline.

Lindberg *et al.* (1986) estimated that atmospheric deposition supplied 40% of the annual requirement for wood production of S and 100% of that for N. Continuing input of acid rain therefore produces imbalances in the ratio of N to cations in tree crowns that result in chlorosis and, if serious enough, death. However, for many forests cause and effect are not easy to link (Binkley 1986; Binkley *et al.* 1989; Markewitz *et al.* 1998). Most forest soils are acidic, so there is the overriding problem of assessing increasing acidification of an already acidic medium; added to this is the difficulty of determining and assessing the sources of acid input.

Two themes show the complexity of the acid rain problem. Markewitz *et al.* (1998) analysed soil samples from the Calhoun Experimental Forest that had been archived over three decades. They concluded that natural processes of acidification (increase in biomass and uptake of $[\text{cations}]_c > [\text{anions}]_c$ by plants, loss of $[\text{cations}]_c > [\text{anions}]_c$ by leaching, increase in soil organic matter, mineralization of N to $NO_3^-$, microbial and root respiration, production of organic acids) accounted for 68% of acidification over the 30 years, with acid deposition accounting for the remaining 32%. With a threefold increase in $SO_2$ emissions for the eastern USA from 1960 to 1980, the capacity of the surface soil to adsorb $SO_4^{2-}$ was saturated but there was strong adsorption below 60 cm. A result of major interest is that, from 1982 to 1990, the concentration of extractable $SO_4^{2-}$ in soils at 35-60 cm decreased significantly. This decrease was equivalent to a loss of charge of $2.2 \text{kmol}_c \text{ha}^{-1} \text{year}^{-1}$, and Markewitz *et al.* (1998) suggest that the decrease in concentration of extractable $SO_4^{2-}$ shows that the soils at Calhoun are responding (or recovering) to the decrease since 1980 in rates of both $SO_2$ emission and $SO_4^{2-}$ deposition.

For the intensively studied Hubbard Brook Experimental Forest, north-eastern USA, inputs of acid rain have caused large losses of cations (particularly $Ca^{2+}$ and $Mg^{2+}$, together with $K^+$ and $Na^+$) in drainage waters (Likens *et al.* 1996). Since 1963, atmospheric emissions have decreased, resulting in decreases in concentrations of $SO_4^{2-}$ and the cations $Ca^{2+}$, $Mg^{2+}$, $Na^+$ and $K^+$ in bulk precipitation. The loss of cations in drainage and the net uptake of cations by the forest together represent an 'ecosystem requirement' that must be balanced by inputs in bulk precipitation, from weathering and from release of cations from exchange sites in the soil. For $Ca^{2+}$, precipitation supplied 29% and net release from soil 63% of this required balance during the period 1940-55. From 1976 to 1993, precipitation supplied only 12% of demand and net release from soil increased to 79%. Likens *et al.* (1996) therefore concluded that even with major reductions in emissions, the response of soil and streamwater chemistry (particularly the acid-base balance) will be significantly delayed relative to that expected from studies of inputs of acids (the bio-geochemistry of sulphur) alone.

## Nitrogen saturation

The second theme is called N saturation, the hypothesis that forests may become 'N saturated' as a result of too much N from acid deposition. Again, this is a concern for the Northern Hemisphere forests, for example in Europe (Skeffington 1990), in North America (McNulty *et al.* 1996) and in Japan (Mitchell *et al.* 1997).

Binkley and Högberg (1997) examined the hypothesis of N saturation in considerable analytical detail for forests in Sweden. While S deposition rates in Sweden have decreased by as much as 50% since 1980, the annual rates are still as high as 25 kg $ha^{-1}$ along the south-west coast, decreasing to <12kg$ha^{-1}$ in the north. These authors also estimated the annual rates of input of N to the south-west coast as 8-12 kg $ha^{-1}$ for $NO_3$—N and 5-15 kg $ha^{-1}$ for $NH_4^+$-N, decreasing in the north of Sweden to <2kg$ha^{-1}$ for both $NO_3$—N and $NH_4^+$-N.

Despite these sometimes high inputs of N, the N is apparently accumulated since in only a few stands near the south-west coast do losses of N by leaching approach inputs of N in deposition. Thus the evidence is that Swedish forests are not N saturated, and forest decline is not linked with any condition of N saturation. The evidence from the application of N fertilizers is uncertain to say the least; some stands respond positively, some negatively and some not at all. Nevertheless, Binkley and Högberg (1997) concluded that 'the vast majority of conifer forests in Sweden still suffer from chronic N deficiency'.

Because of concerns about the acidification of forest soils a number of countries have initiated liming experiments, such as the Höglwald experiment in Germany, where 4tha–$^1$ of dolomitic limestone was applied (Kreutzer & Weiss 1998).

In reviewing these experiments, Binkley and Högberg (1997) concluded that 'the overall effect of liming of forest soils in the Nordic countries has been a general decrease in forest growth (of perhaps 10%)'. A piece of jargon that has emerged from soil acidification and the N-saturation hypothesis is 'vitality fertilization' (or, even worse, 'vitalization fertilization'), a term that really means no more than the application of a fertilizer to ameliorate a deficiency (or, more ornately, to correct a nutrient imbalance). We agree strongly with Binkley and Högberg (1997): 'vitality fertilization' should not be used because it implies lack of vitality where, as in most forests, there is no symptom of decline.

Acid deposition is a problem for the Northern Hemisphere, simply as a consequence of the very much greater density of land area and population in the Northern than in the Southern Hemisphere. While forest decline is caused by acid rain in many forests, in many other forests in both hemispheres decline cannot be linked with high inputs of acid. Wild (1993) summarized a number of causes of decline other than acid deposition.

- Natural acidification processes lead to accumulation of Al and a deficiency of Mg in old trees. This problem is exacerbated by an increase in N supply that causes the development of large crowns, placing demands on Mg. Where Mg is in short supply, Mg is translated from older foliage to developing foliage.
- The increased input of N produces softer or more palatable tissues, resulting in increased insect attack.
- Ozone is directly absorbed by leaves and is extremely damaging to structure and function; decline may therefore be caused by increased inputs of ozone.
- The 1970s and 1980s in Europe were characterized by hot dry summers and cold winters; these conditions might place stress on some species in some areas.
- Poor forest management.

In summary, the 1980s saw the widespread prediction of a general decline of European forests (the concept of *Waldsterben*). By the 1990s this view was greatly modified, particularly by evidence (Kauppi *et al.* 1992, 1995) that the growing stock of European forests increased between 1971 and 1990 by 25% and that growth of the forests increased by 30%. Skelly and Innes (1994) concluded that 'the concept of a general forest decline is untenable', and that the evidence for air pollution causing forest decline 'away from known sources of pollution is extremely limited'.

## TIMBER HARVESTING AND SUSTAINABILITY

### Timber harvesting, nutrient removal and sustained productivity

There have been many studies of the mass of nutrients in the above-ground parts of a wide range of forests, so nutrient removal in harvested timber can be assessed with a degree of certainty. For a range of forests of the eastern USA, Federer *et al.* (1989) calculated that a whole-tree harvest of a mature forest removes about 250kgha$^{-1}$ of N, 20kgha$^{-1}$ of P, 350kgha$^{-1}$ of Ca (up to 600kgha$^{-1}$ for oak-hickory forests), 40kgha$^{-1}$ of Mg and 150 kgha$^{-1}$ of K. Losses of P, Mg and K in nutrient removal and harvest-induced leaching for a single harvest were less than 3% of the total pools, and the loss of N and Ca was less than 8%, except for oak-hickory forests where the loss of Ca was 13-19% (Federer *et al.* 1989). Data for eucalypt forests in Australia (Table 7.5) are similar; losses of N and P due to timber harvesting ranging from 2% when only the stems are harvested up to 5% for whole-tree harvesting.

The magnitude of nutrient loss varies with nutrient, species and harvesting practice, including rotation length, intensity of harvest (debarking or not, stems or whole trees removed) and site preparation for regeneration (particularly whether the logging slash is burned). Judd (1996) simulated nutrient loss due to timber harvesting for a range of harvesting regimes and for two subgenera of the genus *Eucalyptus: Monocalyptus*, which includes the fibrous-barked stringy-barks; and *Symphyomyrtus*, which includes the smooth-barked gums. Judd's model excludes leaching losses. For natural forests with an 80-year rotation, Mg and K are always in positive balance and P always negative. N becomes negative when fire is used to prepare the site for regeneration. Losses of all nutrients increase rapidly as the rotation time decreases. In the extreme comparison, the Ca balance changes over 80 years from +557kgha$^{-1}$ (fibrous-barked natural forest, 80-year rotation, logs debarked on site) to -2168 kgha$^{-1}$ (smooth-barked plantation, 10-year rotation, logs not debarked), a fivefold increase.

This emphasis on the loss of Ca as a potential problem for sustained productivity was also raised by Turner and Lambert (1996) for plantations of *Eucalyptus grandis*, a smooth-barked eucalypt in the subgenus *Symphyomyrtus*. About 60% of the total pool of Ca in a 27-year-old stand was held above ground. Hopmans *et al.* (1993) concluded that Ca and B were the nutrients at greatest risk in managed, natural eucalypt forests in south-eastern Australia. In the study of forests of the eastern USA previously cited, Federer *et al.* (1989) calculated that the present rate of leaching will remove all the Ca in less than 1000 years. If losses due to short-rotation, whole-tree harvesting are added to these leaching losses, 50% of the Ca will be removed in only 120 years.

Clearly this rate of leaching of Ca in the forests of the eastern USA must be recent, otherwise the ecosystem would be entirely depleted. Federer *et al.* (1989) suggest that it has been caused by acid rain. The $H^+$ in solution is replaced by the metallic cations $Ca^{2+}$, $Mg^{2+}$, $K^+$ and $Na^+$. If the input of $H^+$ in acid rain as $H_2SO_4$ and $HNO_3$ was one-tenth the current rate, the metallic cations would be in long-term balance, and this is likely to have been the pre-industrial condition.

By comparison with losses of Ca due to timber harvesting, the losses of N and P (the elements previously and generally considered to be the more limiting and therefore of most likely concern) are relatively small. To put these losses of N and P into a different perspective, we have calculated loss in terms of a change in concentration in the surface 30 cm of soil. For N, the change in concentration of N in the soil over 100 years is a mere 1–2ppm annually (compared with total N about 5000 ppm) and is an order of magnitude less for P, 0.05-0.15 ppm annually (total P about 300 ppm).

In our example of harvesting a eucalypt forest of high productivity on a rotation of 100 years, N balance is restored if the rate of symbiotic $N_2$ fixation is 3.6-6.8 kgha$^{-1}$year$^{-1}$, depending on the intensity of harvesting. *Acacia* species are a natural component of both under-storey and shrub layers in eucalypt forests, and where they grow densely rates of $N_2$ fixation of 20-30 kg ha$^{-1}$year$^{-1}$ have been recorded (Adams & Attiwill 1984a,b; Turvey *et al.* 1984). For most forests, however, the distribution of $N_2$-fixing species is irregular, and that makes the estimation of an areal rate of $N_2$ fixation very difficult. If we are to replace N losses by managing $N_2$-fixing species, we need much more research on managing the rate of $N_2$ fixation by these species relative to the long-term economic value of their N input.

### *Replacing the loss of P: weathering and pools of P*

Resolution of a decrease in P of 0.05-0.15 ppm annually due to harvesting our eucalypt forest on a 100-year rotation is a futile quantitative exercise. Apart from analytical uncertainties in assessing the pool of plant-available P and the rate at which it is sustained over 100 years, there are basic questions that remain unanswered.

1 The rate of rock weathering (see Table 7.5) is a 'ball-park' estimate; how can it be improved?

2 Does P move from lower-class (or less available) reserves, in both surface soil and subsoil, to higher class at the rate of 0.05-0.15 ppm annually?

3 Are tree roots chemically active in obtaining P from lower class reserves? For example, the roots of some forest tree species exude low-molecular-weight organic acids that may play a role in solu-bilizing P (Smith 1976; Fox & Comerford 1990; Fox *et al.* 1990).

### *Replacing nutrient losses in timber harvesting with fertilizers*

The correction of a nutrient deficiency by the addition of an appropriate fertilizer is a simple and relatively economic business. However, compared with agriculture the use of fertilizers in forestry is relatively recent. White and Leaf's (1956) comprehensive bibliography of fertilizer research in forestry contained a mere 700 entries. About 25% of the papers were concerned with nurseries and nursery practice, and some 10% gave attention to 'degraded' forests (also described as 'stunted', 'degenerated', impoverished', etc.) in central Europe (particularly Germany and the former Czechoslovakia) and to the reclamation or 'improvement' of 'useless' land (swamps, bogs, peats, heathlands, moorlands, *Calluna* soils) in the UK, Sweden, Norway, Denmark and Finland and the subsequent establishment of tree plantations.

The 1960s saw the start of extensive use of fertl-izers in forests. Tamm's (1968) historical view of the development of the use of fertilizers in Europe is worth recalling.

- 1865-1900: the recognition, starting with Ebermayer (1876), of nutrient cycling and nutrient demand in forests; almost no experimentation.
- 1900-25: trial-and-error field trials, with little analytical basis.
- 1925-60: scientific foundation for tree nutrition established; many experiments but limited practical application.
- 1960 onward: the start of large-scale application of fertilizers in some forests.

And, indeed, this was a large-scale expansion. For example, the area of fertilized coniferous forest in the north-western and south-eastern USA increased from zero in 1967 to 850 000ha in 1978 (Bengtson 1979). Baule (1973) estimated that the worldwide area of forest to which fertilizer had been applied doubled from 2 million ha in 1970 to 4 million ha in 1973.

While these areas are impressive, we should remember that they are less than 0.1% of the world's forested land (Baule 1973). In short, fertilizers are not used on most of the world's forests. The traditional view of tree nutrition in silviculture of natural forests has always been that 'the loss of nutrient elements through the exploitation of timber is not very serious if the leaves and twigs are left behind in the forest' (Köstler 1956) and that 'while mineral nutrients are as indispensable for trees as for agricultural crops, the cycle [of nutrients] rarely breaks down under rational forest management, and so mineral elements "take care of themselves"' (Baker 1934).

### *Evidence for a decline in productivity due to nutrient removal*

The assertion that the loss of nutrients associated with timber harvesting in natural forests causes a *loss* of productivity is regularly made in the ongoing debate over forests and logging. However, we have not found any definitive quantitative evidence in the literature in support of this assertion, nor are fertilizers used regularly in most of the world's forests used for timber production.

In contrast to the natural forest, fertilizers are used regularly and increasingly in plantation forests around the world. We do not attempt here to cover the many and varied fertilizer prescriptions that have been developed but point to some relatively recent reviews of the use of fertilizer in plantation forests, including Miller (1990) for European forests, Allen *et al.* (1990) for southern pine forests of the USA, Boomsma and Hunter (1990) for *Pinus radiata* plantations in Australia and New Zealand, and Attiwill and Adams (1996) for eucalypt plantations in Australia, New Zealand, Argentina, Chile, Brazil, India, China, Portugal and South Africa.

Plantation forestry very often involves species planted well outside their natural range (e.g. *Pinus* in Australia and New Zealand; *Eucalyptus* in Brazil, South Africa, India, Spain and Portugal), often on soils that are viewed as marginal for most other uses. Nutrient deficiencies and imbalances are usually obvious and can be readily diagnosed and remedied. Furthermore, plantation forestry is more easily managed as a farming process than is the management of a natural forest. In plantation forestry, we manage on rotations of 10 years or so for fibre and on rotations of 15-50 years or so for timber. In natural forests we deal with rotations of 80-150 years. We establish and manage plantations for a specific purpose, whereas we manage a natural forest for all its benefits, such as water production, recreation, aesthetic well-being, conservation of fauna and flora, and timber. The costs of establishing and maintaining plantations are more readily related to a commercial benefit than are the costs of maintaining a natural forest.

Maintaining the productivity of natural forest requires very different scales of perspective, both in time and space, from those that apply in plantation forestry which, in turn, are very different from those that apply to agriculture. In agriculture, we can compare this year's crop with last year's, or with mean yields over the past 10, 20 or 50 years. In plantation forestry, we can compare the productivity of a given compartment with the productivity of the previous rotation. For most natural forests, however, we do not have the records that allow a comparison of the productivity of today's forests with those of 100 years ago.

An often-quoted example of nutrient depletion is the 'second-rotation decline' of *Pinus radiata* plantations in south-eastern Australia. Plantations of pines were established on deep, white, podsolized dune sands that previously supported natural woodlands of no commercial value for timber production. After the first rotation, the litter layer that had accumulated under the pines was heaped and burned along with the logging slash. A serious loss of production between the first and second rotations was clearly recognized by detailed assessment as early as 9.5 years into the second rotation (Keeves 1966). Remedies included both the replacement of nutrients by the addition of fertilizers (Woods 1976) and the conservation of nutrients by the conservation of organic matter between rotations (Squire *et al.* 1985).

We argue strongly against the generalization of productivity decline due to timber harvesting based on the experience in the pine plantations of South Australia, a conclusion also reached in the recent review by Evans (1999), and we argue in particular against the extrapolation of this experience to the natural forest. There are many considerations for the natural forest, including rotation length, intensity of harvest, the appropriateness of harvesting methods and silvicultural treatments relative to the ecology of natural regeneration, and the influence of other factors (e.g. Ca depletion due to increasing acid rain in some forests of eastern USA; Federer *et al.* 1989).

### *Nutrient demand in the natural forest*

The total cycle of a nutrient in a forest is the sum of all the fluxes of the nutrient between plant and soil *plus* that amount of nutrient needed for new growth. The total cycle is calculated as the sum of the amounts of nutrient: (i) cycled from plant to soil in litterfall; (ii) recycled within the plant; (iii) in inputs less that in outputs; and (iv) taken up from soil reserves. For a mature forest, the percentage contribution of litterfall to the total annual cycle decreases in the order Ca > Mg > N > K > P, canopy leaching K > Ca = Mg > N > P, internal redistribution *P* > N > K > Mg > Ca, inputs-outputs N = Ca > Mg > K = P, and net supply from soil reserves K > *P* > N > Mg > Ca. For all the nutrients discussed here, annual uptake by the mature forest from soil reserves accounts for less than 20% of the total annual demand.

Of course, the relative contributions of nutrient cycling pathways to the total cycle must change with age of the forest. Attiwill (1980) defined three stages of forest growth and nutrient cycling, and Miller (1981) related the response of forests to fertilizers within essentially the same three stages.

1 Up to 20 years or so, the forest builds its productive system; the major proportion of NPP is used in building the photosynthetic crown and the systems for transport of water, nutrients and assimilates. This period is dominated by uptake from soil reserves, and Miller writes that this is the stage 'when responses to a wide range of fertilizer [additions] can be expected'.
2 The rate of increase of the productive system is slowed. An increasing proportion of NPP is stored as physiologically dead heartwood, internal recycling accounts for an increasing proportion of the total cycle of the physiologically mobile nutrients, and the dependence on uptake from soil reserves decreases. A response to fertilizer might only be expected if the canopy is thinned.
3 The trees reach maturity and the rate of biomass increase approaches zero. The major proportion of NPP is now shed as litter. Nutrients (particularly N and P) may be immobilized in the litter; additions of N and P may speed up decomposition but will have little effect on growth of the trees (e.g. Gholz & Fisher 1984).

Does the addition of fertilizer increase the carrying capacity of a forest? For agricultural crops, forestry where productivity has been restored on degraded land, and plantations that have been established on previously unproductive (in terms of timber production) land, limiting factors are recognizable and some of them can be easily remedied. For example, on some of the sandy soils in Australia, the success of agricultural and tree crops depends on additions of superphosphate and trace elements such as Cu and Zn. However, the productivity of ecosystems (including agricultural and forest crops) is up to an order of magnitude less than so-called 'potential' production (e.g. Jarvis 1981; Axelsson 1983). Remedies such as draining, ripping the subsoil, mounding, adding fertilizers and irrigating overcome some, but by no means all, the limitations.

However, for the natural forest and many plantations, it seems that an increase in carrying capacity due solely to the addition of fertilizers is unlikely. Nor is it likely that fertilizers will be used extensively to increase the rate of production. The efficiency with which trees use nutrients added as fertilizer is low, and growth responses are short-lived (Cole *et al.* 1990). It is more likely that fertilizers will be used at relatively low rates to maintain productivity and essential ecosystem processes such as mineralization (Gessel & Atkinson 1984; Gholz & Fisher 1984).

## Timber harvesting, soil compaction and sustainability

Timber harvesting and logging equipment compact the soil and therefore change the physical properties of soil. Compaction is due to a decrease in macropores to a depth of 30 cm or so, resulting in increases in bulk density, water-filled porosity, available water and soil strength, and in decreases in sorptivity, aeration porosity and hydraulic conductivity (Rab 1994, 1996; Herbauts *et al.* 1996; Huang *et al.* 1996; Woodward 1996; Page-Dumroese *et al.* 1998; Stone & Elioff 1998; Merino & Edeso 1999).

Despite the magnitude of some of these changes (e.g. an increase in bulk density in the surface 0-10 cm of soil of 20-40%; Rab 1996), the effects of compaction on plant growth are varied. Where disturbance by machinery is greatest (e.g. skid trails and log landings) the effects of compaction are compounded by the removal of organic layers and topsoil and by the exposure of subsoil. Among tropical forests, for example, height growth of mahogany was unaffected by soil compaction (Whitman *et al.* 1997), whereas growth of a number of species in Sabah, Malaysia, and in Amazonian Ecuador was reduced by compaction but responded more to the addition of fertilizers than to a range of treatments designed to ameliorate the effects of compaction (Nussbaum *et al.* 1995; Woodward 1996). Soil compaction may affect root growth more than shoot growth (Page-Dumroese *et al.* 1998).

Given that the snig tracks and log landings are properly revegetated, the physical properties of soils recover to precompaction conditions within 5-40 years, depending on soil type, organic content and intensity of compaction (Dickerson 1976; Greacen & Sands 1980; Wingate-Hill & Jakobsen 1982). Thus one could conclude that where recovery time is less than one rotation, the effects of compaction *per se* will not be cumulative (Worrell & Hampson 1997). However, the problem of loss of topsoil through bad forestry practice seems to be of greater and longer-term significance than the problem of compaction alone. Restoring the fertility and physical structure of topsoil will take centuries rather than decades.

As well as direct effects of compaction and removal or erosion of topsoil on the subsequent growth of trees, there may be direct effects on diversity. The topsoil is a store of seed of some plant species and of vegetative propagules for others. Significant decreases in number or abundance of plant species have been observed after logging in aspen forests in the northern Lake States region, USA (Stone & Elioff 1998), in rain forest in Sabah, Malaysia (Nussbaum *et al.* 1995), and in mountain ash forest in Victoria, Australia (Ough & Ross 1992; Ough & Murphy 1997).

In summary, conservative management of the soil to minimize erosion and compaction is fundamental to sound land management; it is as essential in forestry as it is in agriculture. However, the nature of soil compaction in forests differs from that in agriculture (Greacen & Sands 1980). The mass of the trees, the forces generated by wind and the increasing size of roots all lead to soil compaction. Equipment used in timber harvesting is heavy, and together with the movement of large logs exerts great pressure on soils. Finally, equipment used in timber harvesting does not traverse the land as uniformly as that used in agriculture, so that heterogeneity of compaction and soil disturbance is greater in forestry than in agriculture (Greacen & Sands 1980).

Furthermore, timber harvesting generally requires the construction of roads and logging tracks to reach new logging areas. These new roads and tracks are particularly prone to erosion. Siltation of streams due to roading plus logging may be orders of magnitude greater than siltation due to logging alone (Grayson *et al.* 1993; Dargavel *et al.* 1995). Conservative management of soil in forest operations must therefore be all-embracing. Roads, skid trails and log landings must be carefully planned: steep slopes must be avoided, river crossings minimized, cross-drains created to disperse water, buffer zones established and maintained, and revegetation hastened (Lamb 1986). All these requirements can be met through strict adherence to, and monitoring of, codes of forest practice.

Assessment of the physical condition of a coupe after logging, especially the amount of soil disturbance, is targeted through sustainability indicators developed by the Working Group (1995) for the Montreal Process. Where fire is an essential component of forest ecosystems, these sustainability indicators must be sensibly interpreted. For example, *Eucalyptus regnans* (mountain ash) in Victoria, Australia, regenerates in nature only after bushfire, and the ash-bed effect (the burning of all shrub, ground and litter layers and heating of surface soil) is essential to germination, survival and growth (Ashton & Attiwill 1994; Chambers & Attiwill 1994; Ashton & Kelliher 1996). In commercial forestry operations, the trees are clearfelled and the logging slash is burned to prepare the seed bed. Subsequent seedling growth is always significantly greater on burned soils than on unburned soils, including those where a seed bed has been mechanically prepared (Van Der Meer *et al.* 1999). Sustainability indicators that emphasize retention of litter layers to protect topsoil (e.g. Rab 1999) will therefore have little basis in ecological processes when fire is involved.

# 8

# Models for Pure and Mixed Forests

## INTRODUCTION

The previous chapter showed how forest growth is the sum of their species interactions with environment as mediated through physiological processes, and the dynamic nature of the responses of trees and stands. Long before forest scientists understood even some of the more basic processes governing the way trees grow, considerable empirical knowledge had accumulated through observation to quantify tree growth. While based on observation and not first causes, it is no less a contribution to understanding how trees grow and what affects them by looking at the way they respond in the 'field' or more properly the 'forest'!

This chapter reviews our understanding of tree and forest stand behaviour gained from empiricism (what has been observed), through quantifying systematically such observations (the yield table) to process-based modelling of both pure and mixed forest stands. For a deeper understanding of the introduced model types, the bibliography offers the most important references. A model's objective and existing knowledge about the observed system determine how complex the model approach has to be. Single-tree models, ecophysiological-based gap models and hybrid models are of particular interest for forest management as they are suitable across many stand types and forest conditions.

Forest growth models aggregate knowledge of individual processes of forest growth to predict stand or whole system functioning. Forest ecosystems may be modelled with varying degrees of temporal and spatial resolution. The time-scale may range from seconds to millenia, while the spatial scale may encompass anything from cells and mineral surfaces to continents. The slow processes on large spatial scales fix the boundary for quicker processes on smaller scales. Conversely, the rapid and spatially bounded processes determine the processes on higher levels. Model approaches that take into consideration these feedback loops between the different system levels can provide important contributions to system understanding as well as to management decision support. At our present state of system knowledge, single-tree and stand models, which model processes on temporal scales from a year to a century and on spatial scales from tree to stand level, best fullfil the demands of forest management. They model the stand dynamic on the basis of the classical growth and yield variables, like diameter, height, crown length, etc. Process models have a higher spatial and temporal resolution and approximate to the classical variables by scaling-up. However, the behaviour of the whole system can be more than the sum of the underlying processes. Forest sucession and biome shift models become an important tool for global change research in forest ecosystems.

## PATTERNS AND DYNAMICS OF GROWTH: EMPIRICAL OBSERVATIONS

### Periodicity and pattern in individual tree growth

Tree growth is periodic over short time-spans and follows a definite pattern in the long term. In temperate countries growth is overwhelmingly determined by the season, with growth confined to a period of a few weeks to perhaps several months in any one year while warmth and moisture are adequate. Even in the moist tropics, with year-round favourable growing conditions, trees show periodicity under a measure of genetic control as recorded, for example, from measurements of multinodal tropical pines such as *Pinus caribaea* (Chudnoff & Geary 1973) and fast-growing hardwoods such as *Cordia alliodora* (Blake *et al.* 1976).

As well as periodicity in one year, trees exhibit a strong pattern of growth during their life. In relation to age the pattern is typically sigmoid, whether recording total height, diameter or volume over time.

## Dynamics of stand growth

The dynamic nature of stand growth is a little less obvious than for a single tree, since superficially a stand seems to be a collection of individual trees the relations of which appear constant. That this is not so is seen by considering a small tree in a mature even-aged stand. At regeneration (or planting) this small tree will have been about the same size as its neighbours but subsequently will have competed less successfully for light, nutrients or moisture to become inferior to its neighbours. As a stand develops any apparent uniformity disappears, with some trees developing vigorously and neighbours becoming suppressed, moribund and even dying. Thus the stand is continually changing - it is dynamic. It is not only a collection of individual trees, each with its own genetic potential for using the site, but also a collection of trees that interact and compete with one another.

### *Seedlings and regeneration*

From the outset trees appear to grow at different rates. Even in new plantations of well-spaced trees of one species, and long before onset of between-tree competition, the growth of individual trees is not identical. Moreover, there is often no correlation between initial postgermina-tion vigour and subsequent growth rate. Small seedlings, or plants from the forest nursery, do not necessarily lead to small slow-growing trees (see illustration of this for young *Pinus caribaea* in Evans 1992, p. 221).

### *Between-tree competition*

As trees grow they eventually begin to compete with their neighbours. This has both an above-ground component, mainly competition for light, and a below-ground element in terms of root competition for nutrients and moisture. Such competition is most readily seen in suppression of side branches on the lower crown and a measurable impact on diameter increment. The timing of the onset of between-tree competition depends on distance between trees, but is usually first measurable in forest stands from about the time of canopy closure. The principal exception is in arid climates where competition between root systems for moisture greatly exceeds that for light, and trees and shrubs remain widely dispersed and rarely in contact above ground.

*Differentiation into crown classes*

As competition begins, growth of some trees slows more than others. These tend to be the smaller trees at the time of canopy closure and the competition reinforces their inferior status. Once dominated, few trees can recover unless a gap develops owing to death of a neighbour. Thus as a stand develops, a range of tree sizes emerges and, traditionally, these are classified into different crown classes according to the tree's relative position in the canopy.

### Interventions and manipulations

These empirical relationships of how trees are observed to grow and how they develop and interact in a stand provide the basis for manipulating their behaviour. Adjusting spacing between trees, both when planted or through thinnings, or influencing the balance of a mix of species in a stand all impact on how a stand develops, on how the increment is distributed and hence on the composition of tree types, trees sizes and the total of woody growth that will result.

In summary, densely stocked forest leads to high volumes per unit area but small mean tree size compared with less well-stocked forest on a similar site and of the same age. In the latter case there will be fewer but larger diameter trees, although total volume of timber may be somewhat reduced. This allows forests to be managed in different ways to yield different assortments of products. How to model these relationships and outcomes in detail, beyond the purely empirical, forms the bulk of this chapter.

## GROWTH MODELS

The history of forest growth models is not simply characterized by the development of continuously improved models replacing former inferior ones. Instead, different model types with diverse objectives and concepts were developed simultaneously. The objectives and structure of a model reflect the state of the respective research area at its time and document the contemporary approach to forest growth prediction. The history of growth modelling thus also documents the advancement of knowledge in the science of forest growth.

Beginning with yield tables for large regions as a basis for taxation and planning (such as those by Schwappach 1893 and Wiedemann 1932, 1939a,b, 1942), model development led to regional yield tables and site-specific yield tables and culminated in the construction of growth simulators for the evaluation of stand development under different management schemes. Vanclay (1994) provided an overview about growth and yield management models and their application to mixed tropical forests. The 1980s brought a new trend with the development of ecophysiological models, which give insight into the complex causal relationships in forest growth and predict growth processes under various ecological conditions. The emphasis in model research has shifted towards ecophysiological models and away from models aimed only at providing growth and yield information for forest management. These models attempt to simulate forest growth on the basis of fundamental ecophysiological processes. The scientific value of ecophysiological models cannot be overrated; however, they will not be applied in forest management for the next few years as they are in many ways not yet sufficiently validated. Also, input and output variables do not yet meet the demand of forest management practice.

A major change has taken place in model conception, i.e. the understanding of forest growth on which the model is based. The tables by Weise (1880), Schwappach and Wiedemann resulted from a purely descriptive analysis of sample area data in the form of total and mean values of observed processes of stand development. These descriptions were later combined with theoretical model concepts that also considered natural growth relationships and causal relations as far as they were known at the time. For example, yield tables for mixed stands of pine and beech created by Bonnemann (1939) characterize growth of beech in the middle and lower storey by mean values. The FOREST model of Ek and Monserud (1974) controls increment behaviour of lower-storey trees by geometrical competition indices, and the ecophysiological growth models of Bossel (1994), Mäkelä and

Hari (1986) and Mohren (1987) derive increment behaviour of lower-storey trees from light availability and performance in terms of photosynthesis.

The change in model objectives and concepts is closely related to a change in quality of the information generated. Pure management models aim at reliable prediction of forest yield values that are crucial for planning and control in forest management, e.g. height and diameter increment and associated economic value. Ecophysiological models aim at biomass development, nutrient input and loss, etc.; variables relevant to forest management are only of secondary importance in these models. For future planning in modern forestry, models meeting the information demands of ecology as well as of economy will gain in importance. Ecophysiological models and stand management models can give specific decision support. Ecological and socioeconomic conditions define the framework and thus the 'foundations' for management decisions. Ecophysiological models can support the ecological elements of the framework, for example the effect of site conditions, species mixture and thinning variants on critical loads, water quality or acidification. Stand treatments of interest can be judged in this way as ecologically acceptable or unacceptable. Management models help to optimize the path from starting point to objective via the given framework, for example they support the decision between different thinning and pruning strategies.

With the shift from tree and stand management models with low resolution to more complex ecophysiological models, different source data are needed for model construction and for the determination of model parameters. Standard datasets derived from research sample plots (diameter, height, etc.) were used for the development of stand growth models for applied forestry. For the construction of single-tree models, additional data are required (crown dimension, tree position, etc.). The transition to ecophysiological models requires an additional database that can only be provided by broadening experimental concepts and cooperation with neighbouring disciplines.

Models are always an abstraction of reality and are greatly influenced by the modeller's knowledge and perception of nature.

This applies to the construction of yield tables as well as for eco-physiological models.

## Stand growth models based on mean stand variables

With a history of over 250 years, yield tables for pure stands may be considered the oldest growth models in forestry science and forest management. They are representations of stand growth within defined rotation periods and are based on a series of measurements of diameter, height, biomass, etc. reaching far back into the past. From the late eighteenth to the middle of the nineteenth century, German scientists such as Paulsen (1795), von Cotta (1821), R. Hartig (1868), Th. Hartig (1847), G.L. Hartig (1795), Heyer (1852), Hundeshagen (1825) and Judeich (1871) created the first generation of yield tables based on a restricted dataset. These original yield tables soon revealed great gaps in scientific knowledge. A series of long-term data-collection campaigns on experimental areas was therefore started. This was the birth of a unique network of long-term experimental plots in Europe that is still under survey.

The second generation of yield tables, initiated towards the end of the nineteenth century and continued into the 1950s, follows uniform construction principles proposed by the Association of Forestry Research Stations (the predecessor organization of the International Union of Forest Research Organizations (IUFRO)), in 1874 and 1888 and has a solid empirical basis. The list of protagonists involved in this work includes Weise (1880), von Guttenberg (1915), Zimmerle (1952), Vanselow (1951), Krenn (1946), Grundner (1913) and, in particular, Schwappach (1893), Wiede-mann (1932) and Schober (1967), who designed yield tables that were conceptually related and is still being used to this day. A brilliant example of their work is the yield tables for European beech. In the 1930s and 1940s the first models of mixed stands were contructed under the direction of Wiedemann. Data from some 200 experimental areas established by the Prussian Research Station led to the yield tables for even-aged mixed stands of pine and beech (Bonnemann 1939), spruce and beech (Wiedemann 1942), pine and spruce (Christmann 1949), and oak and beech (Wiedemann 1939a). The Second World War prevented

Wiedemann from bringing the development of yield tables for uneven-aged pure and mixed stands to an end, but his studies initiated systematic research on mixed stands. Yield tables for mixed stands of this generation were never consistently used in forestry practice as they were restricted to specific site conditions, intermingling patterns and age structures.

Yield tables developed by Gehrhardt (1909, 1923) in the 1920s effected a transition from purely empirical models to models based on theoretical principles and biometric formulae and led to a third generation of yield tables. These models were designed by, among others, Assmann and Franz (1963), Hamilton and Christie (1973, 1974), Vuokila (1966), Schmidt (1971) and Lembcke *et al.* (1975), and at their core is a flexible system of functional equations. These functional equations are based as far as possible on natural growth relationships and are generally parameterized by means of statistical methods. The biometric models are usually transferred into computer programs and predict expected stand development for different spectra of yield and site classes. A wealth of data were available for the construction of these models and processed with modern statistical methods.

Since the 1960s a fourth generation of yield table models has been created, i.e. the stand growth simulations of Franz (1968), Hoyer (1975), Hradetzky (1972), Bruce *et al.* (1977) and Curtis *et al.* (1981, 1982), which simulate expected stand development under given growth conditions for different stem numbers at stand establishment and for different tending regimes. They describe stand development at different sites and for varying treatments and varying numbers of trees at the time of establishment. Expected stand development under given growth conditions is simulated by means of computer programs and controlled by systems of suitable functions forming the core of the growth simulator. All information available on forest growth is synthesized into a complex biometric model that simulates stand development for a wide range of possible management alternatives and summarizes the results in tabular form similar to yield tables. These yield tables reflect the stand dynamic for a wide range of imaginable management scenarios. While table and model were identical for the yield tables of earlier generations, simulator-created yield tables now describe just one of many potentially computable stand development courses.

Despite a number of drawbacks, yield tables still form the backbone of sustainable forest management planning. When computing capacities and available data for model construction increased and with the rising demand for information in forestry, mean-value and sum-orientated growth models and yield tables were increasingly replaced by stand-orientated growth models, predicting stem number frequencies, and by single-tree growth models. Prodan (1965, p. 605) commented on the significance of yield tables in the context of silviculture and forest sciences: 'Undoubtedly, yield tables are still the most colossal positive advance achieved in forest science research. The realization that yield tables may no longer be used in the future except for more or less comparative purposes in no way detracts from this achievement'.

## Stand-orientated management models predicting stem number frequency

With the transition towards new intensive treatment concepts, the demand for information in forestry has changed the emphasis from mean stand values towards single-tree dimensions of selected parts of a stand. This changed demand for information resulted in the 1960s in the creation of the first growth models, which enabled prediction of mean stand values as well as frequencies of single-tree dimensions. Until then, a stand served as the usual information unit on which all predictions were based; these predictions were now strengthened by statements about stem number frequencies in diameter classes, which are needed for precise prediction of assortment yield and value of a stand. Depending on their concept and construction, stand-orientated growth models predicting stem number frequency are classified into differential equation models, distribution prediction models and stochastic evolution models.

Many natural processes in various disciplines of the natural sciences can be described by differential equations. Examples are the differential equations formulating change of yield descriptors for diameter classes of a stand, i.e. change of stem number, basal area and growing stock, depending on current yield state values. Stand development then results from the numerical solution of the differential equations. In the 1960s and 1970s, Buckman (1962), Clutter (1963), Leary (1970), Moser (1972, 1974) and Pienaar and Turn-bull (1973) developed stand-orientated growth models based on differential equations.

In the mid 1960s, Clutter and Bennett (1965) proposed a completely new approach to stand growth modelling. They characterized the condition of a tree population by its diameter and height distribution and described stand development by extrapolation of these frequency distributions. The precision of such models is decisively determined by the flexibility of the distribution type on which it is based. The suitability of different distribution types, for example beta, gamma, lognormal, Weibull or Johnson, has to be assessed individually. Compared with those reviewed earlier, in these models stand development is not controlled by the age function of the individual yield descriptors but by the parameters of the underlying frequency distribution. Models of this type were initially constructed by Clutter and Bennett for North American spruce stands and further developed by McGee and Della-Bianca (1967), Burkhart and Strub (1974), Bailey (1973) and Feduccia *et al.* (1979).

The term 'evolution models' for stochastic growth models is derived from the fact that in these models stand development evolves from an initial frequency distribution, for example from a diameter distribution known from forest inventory. Thus these models, like distribution prediction models, predict frequencies of single-stem dimensions. However, the mechanism accounting for the extrapolation is based on a Markov process, giving the transition probability for the shift between the diameter classes. Stochastic growth models were introduced to forestry science with the pioneering investigations by Suzuki (1971, 1983), and they continue to be linked to his name today. His growth models, for Japanese *Chamaecyparis* pure stands for example, have been consistently developed by Sloboda (1976) and his team since the mid 1970s; they are mainly interested in adapting the models, which are orientated to Japanese conditions, to the issues of German forestry and in model validation based on permanent test plot data. Stand-orientated growth models based on stochastic processes have been developed by Bruner and Moser (1973) and Stephens and Waggoner (1970) also for mixed stands.

## Single-tree orientated management models

Single-tree models describe the stand as a mosaic of single trees and model individual growth and interactions with or without consideration of tree position. This has paved the way for the design of models of pure and mixed stands of all age structures and intermingling patterns. An equation system that controls growth behaviour of single trees depending on their constellation within the stand is the central module of all single-tree models. Position-independent or position-dependent competition indices are used to quantify the spatial growth constellation of each tree and to predict its increment of height, diameter, etc. in the following period. Compared with stand-orientated growth models based on mean stand descriptors and those predicting stem number frequencies, single-tree models work on higher resolution. The information unit in singletree models is the individual tree. However, results of lower-resolution models, for example mean tree development or diameter frequency distributions, can also be derived from single-tree model results by integration. Information about stand growth then results from summarizing and aggregating each individual single-tree development for a given growth period. Recent single-tree models are programmed to enable the user to influence a simulation run interactively. This allows stand development to be followed step by step during the simulation and permits the user to specify other factors (e.g. thinning or influence of disturbance) at any time during the simulation process, thus influencing or diverting the current course of stand development.

After parameters for the control of the singletree model have been set, tree characteristics at the beginning of the prediction phase for the test area to be investigated are fed into the computer as initial values for the simulation. This tree list can contain data on tree species, stem dimensions, crown morphology, stem position and other data about the stand individuals. These data usually originate from single-tree-based inventories of indicator plots.

Starting with these initial values, change (e.g. mortality or development of diameter, height or crowns) for all stand members depending on individual growth conditions is predicted using an appropriate control function; this is done for a first growth period, for example 5 years. Once the tree list has been processed, change of growth conditions (e.g. due to thinning or disturbance) can be specified prior to continuing to the next increment period. This will now influence single-tree growth in the following period. The modified state values of all trees resulting at the end of the first growth period also represent the initial values for the second growth period. These values are repeatedly extrapolated in every simulation cycle and interim results are given. The simulation continues until the envisaged prediction period has been completed step by step. In most models, time steps are 5 years, sometimes only 1 or 2 years. By removing single trees during a simulation run, the growth constellation and growth behaviour of the remaining individuals change in the next growth period. Growth reaction of the stand is thus explained by the reactions of all single trees to this intervention. By relating stand development back to growth behaviour of single trees and by modelling single-tree dynamics depending on growth constellation within the stand, single-tree models, after being initialized accordingly, enable evaluation of a wide range of treatment programs.

The first single-tree model was developed for pure Douglas fir stands by Newnham (1964). It was followed by the development of models for pure stands by Arney (1972), Bella (1970) and Mitchell (1969, 1975) and colleagues. In the mid 1970s, Ek and Monserud applied the construction principles for single-tree orientated growth models for pure stands to uneven-aged pure and mixed stands (Ek & Monserud 1974; Monserud 1975). Munro (1974) distinguished distance-dependent and distance-independent single-tree models, the former being able to refer to data about stem position and stem distance for the control of single-tree growth. The worldwide bibliography of single-tree growth models compiled by Ek and Dudek (1980) lists more than 40 different single-tree models, which are grouped into 20 distance-dependent and 20 distance-independent models. Single-tree models developed since the 1980s (Wykoff *et al.* 1982; van Deusen & Biging 1985; Wensel & Koehler 1985; Pretzsch 1992, 1998, 1999; Sterba *et al.*

1995; Nagel 1996) in many ways go back to the methodological bases of their predecessors; however, owing to the rapidly improving technology of modern computers they are far more user-friendly than older single-tree models.

## Ecophysiological growth models

All the models mentioned above rely on growth and yield data from long-term observation plots and hence have the advantage of being validated empirically. However, there is a drawback to historically deduced data in as much as growth conditions undergo changes, and reaction patterns from the past cannot simply be projected into the future. In the 1970s, model research was pointed in a new direction with the creation of high-resolution ecophysiological process models, which account for metabolism, organ formation, assimilation and respiration as well as biochemical and soil chemistry reactions. Pioneers of the ecophysiological process model for forest stands are Bossel (1994), Mäkelä & Hari (1986) and Mohren (1987). The term 'process model' is slightly misleading in the sense that all forest growth models describe processes. Only the temporal and spatial scales of modelled processes become more detailed and accurate in the transition from yield table models via single-tree management models and succession models to growth models based on ecophysiological data.

The development of modern process models begins with a systems analysis and the selection of characteristic system components. A system to be analysed and modelled is first described using methods of systems analysis. Results of this description can be transferred into a system diagram. The description breaks the system down into system components characteristic for all biological systems and identified by different symbols in the system diagram. By system parameters we mean those that remain constant during the lifetime of the system. Exogenous parameters are variables that control the system but which cannot be influenced by the system, e.g. stress caused by air pollutants. State variables are the actual output value of the model; their current values reflect the system's state. Important state variables in stand models are accumulated carbon quantities in needles, branches, stem and roots. The initial values of the state variables give the starting values of a system and thus crucially influence its further development. In a growth model for example, stem number and initial stand structure have to be specified as initial values. The rate of change of the state variables controls change, i.e. input and output of state variables. Examples are mortality rates or respiration rates, which control the change of the carbon quantities accumulated in the different components. Intermediary variables change simultaneously with the state variables and feed back into the system. The system components are indicated in the system diagram with different symbols and their interrelations are identified by arrows.

The model thus outlined is transferred into a mathematical model and subsequently into a computer program. For this, the system components and links are described by mathematical or logical relationships. Once the complete model is constructed, the causal relations implemented are parameterized. The system behaviour can be simulated with the developed computer programs. All suitable information known about the system is therefore consolidated in the system components and the system structure. The process of system analysis and model development concludes in the validation of the final model. For validation, i.e. testing if the causal relations assumed in the model realistically reflect growth of stands or single trees, empirical yield data can be used. If necessary, individual model assumptions are corrected or model parts revised.

A vastly improved understanding of ecophysiological processes in forest ecosystems paved the way for this model approach and it was the actual modelling of these processes that provided an idea of the functioning of the overall system. A further impetus to process model development was the need to understand and predict the reactions of forest ecosystems to an increasing number of adverse effects, such as industrial emissions, rise in atmospheric $CO_2$ and climate change. In the context of environmental instability, high-resolution and accurately detailed process models are certainly the ideal approach for understanding and predicting forest ecosystem behaviour. However, there are particular constraints in developing and applying process models due to considerable gaps in our knowledge of part processes in assimilation organs and in the soil.

Also, the scaling-up of part processes to the behaviour of the overall system is still largely unresolved. Moreover, the introduction of process models still requires intensive research and extremely high-powered computers that are only rarely available in practice. To date, process models are therefore primarily research instruments rather than forest management planning tools.

## Gap models and biome shift models

In the view of modern theoretical ecology, a spatially extensive system is composed of mosaiclike subunits and can be studied by analysing these subunits. Watt (1925, 1947), Bormann and Likens (1979) and others transferred this view of extensive ecosystems to the study and model representation of the growth dynamics of pure and mixed stands. This laid the foundations for the concept of gap models suitable to predict succession. According to this concept, a forest stand is an aggregation of gaps. The size of these gaps corresponds to the extent of a potential crown area of a dominant tree or tree group (areas of 0.04-0.08 ha). The actual information unit is the tree group in the gap; stand development results as the sum of the total spectrum of contributing gaps. Gap models imply that forest development in a gap occurs in a fixed cycle: A gap results from exploitation or death of a dominant tree, and thus the growth conditions of under-storey trees improve and natural regeneration occurs. Growing trees successively close the gap and a new overstorey develops. The cycle is repeated with further losses of dominant trees. Growth models using this approach were predominantly employed for investigations of competition and succession in semi-natural stands.

Gap models, such as those designed by Shugart (1984), Pastor and Post (1985), Aber and Melillo (1982) and Leemans and Prentice (1989), are primarily aimed at mixed stands. While in the models described above increment-determining factors have effects on stands or individuals respectively, gap models describe tree growth that depends on growth conditions in the individual gap. Gap models simulate growth dynamics for single trees or tree classes in a gap; it is therefore possible to generate information about the development of diameter, height and volume of single trees as well as stands.

However, regarding input and output variables they are less dependent on information available from, or required by, forestry practice; rather, they aim at predicting long-term succession in natural forest stands and the effects of altered growth conditions. The FORMIX2 model (Bossel & Krieger 1994) for virgin and logged Malaysian lowland dipterocarp forests is an example of an ecophysiological-based gap model with output variables that is useful as decision support in forest management.

Biome shift models, such as those of Box and Meentemeyer (1991) and Prentice *et al.* (1992), establish statistical relationships between regional climate and vegetation type. Based on relevant climatic conditions, the nature of potential biomes, i.e. communities, may be predicted on a regional and even global scale. Of all the models under discussion, these are the ones that provide the highest aggregation of data on vegetation development and forest growth. They have therefore gained increasing importance in research on global change.

## Hybrid models for forest management

The transfer of specific components of ecophysio-logical models (based on solid process knowledge) into stand or single-tree management models (based on long-term experimental plots and increment series) leads to what Kimmins called 'hybrid growth models'. Models of this type were constructed by, among others, Botkin *et al.* (1972) and Kimmins (1993). Their objective is to make the best possible use of the newly acquired knowledge of ecophysiological processes combined with historical increment observations to assist in forest planning and management. On account of the implemented relationship between site conditions and species-specific growth, they can be used for pure and mixed stands. In the past 100 years mixed stands have gradually become the focus of forest research, particularly on account of studies by Gayer (1886), Wiedemann (1939b) and Assmann (1961), but to this day growth models for mixed stands are scarcely used as quantitative planning tools.

Only very recently have models created by Kolström (1993), Nagel (1996), Pretzsch (1992), Pukkala (1987) and Sterba *et al.* (1995) found use in forestry practice for planning work in pure and mixed stands. These are in effect site-sensitive single-tree models constructed from a broad base of ecophysiological and growth and yield data. Version 2.2 of the SILVA model, developed in Germany for pure and mixed stands, belongs to the category of hybrid models (Pretzsch 1992; Pretzsch & Kahn 1996; Kahn & Pretzsch 1997) and may be used as an example to explain the functional principles underlying this approach.

## Management model SILVA 2.2 for pure and mixed stands

SILVA reflects the spatial and dynamic character of mixed-stand systems in as much as it models spatial stand structures at 5-year intervals. This permits the recording of the individual growth constellation of every tree and the control of tree increment in relation to growth constellation and the original dimensions of the tree (Fig. 9.8). The external variables determining tree increment and stand structure are treatment, risk and site factors. The model simulates the effects that tending, thinning, regeneration and natural hazards such as storms and wind have on the stand dynamic.

The feedback loop, stand structure —> tree growth —> state of tree —> stand structure, forms the backbone of the model. The step-by-step modelling of the growth of all individual trees via differential equation systems provides information about the development of assortment yield, financial yield, stand structure, stability and diversity of the stand over and above the data, required in yield calculations, on height, diameter at breast height, number of stems, etc.

Input and output data used in the model correspond to the data available from, or required in, forestry practice, for example only site variables available on a large scale are considered. With models of this type a weighting between yield-related, socioeconomic and ecological aspects of stand development in pure and mixed stands becomes possible. Parametrization relies on yield and site characteristics of pure and mixed stands that have been under observation for over 100 years.

The position-dependent individual tree model SILVA 2.2 breaks down forest stands into a mosaic of individual trees and reproduces their interactions as a space-time system. It can therefore be used for pure and mixed stands of all age combinations. Primarily it is designed to assist in the decision-making processes in forest management. Based on scenario calculations SILVA 2.2 is able to predict the effects of site conditions, silvicultural treatment and stand structure on stand development, and therefore also serves as a research instrument.

A first model element reflects the relationship between site conditions and growth potential and aims at adapting the increment functions in the model to actual observed site conditions. With the aid of nine site factors reflecting nutritional, water and temperature conditions, the parameters of the growth functions are determined in a two-stage process (Kahn 1994). The stand structure generator STRUGEN facilitates the large-scale use for position-dependent individual tree growth models. The generator converts verbal characterizations as commonly used in forestry practice (e.g. mixture in small clusters, single tree mixture, row mixture) into a concrete initial stand structure with which the growth model can subsequently commence its forecasting run (Pretzsch 1997).

The three-dimensional structure module uses tree attributes such as stem position, tree height, diameter, crown length, crown diameter and species-related crown form to construct a spatial model of the stand in question. The thinning model is also based on individual trees and can model a wide spectrum of treatment programs (Kahn 1995). The core of the thinning model is a fuzzy logic controller. In the simulation studies described below the thinning model simulates various thinning methods (thinning from below and selective thinning) and thinning intensities (slight, moderate and heavy). The competition model employs the light-cone method (Pretzsch 1992) and calculates a competition index for every tree on the basis of the three-dimensional stand model. The allocation model controls the development of individual stand elements. Tree diameter at height 1.3 m, tree height, crown diameter, height of crown base, crown shape and survival status are controlled, at 5-year intervals, in relation to site conditions and interspecific and intraspecific competition. Finally classical yield information on stand and single-tree level for the prognosis period are compiled in listings and graphs. Additional information on stem quality, assortment and financial yield complete the growth and yield characteristic. At every stage of the simulation run, a program routine for structural analysis calculates a vector of structural indices that serve as indicators for habitat and species diversity and form a link to the ecological assessment of forest stands.

The algorithmic sequence for predicting forest development comprises the following steps. Step 1 is the input of data on the initial structure and site conditions of the monitored stand. In step 2, the parameters of the growth functions are adapted to actual site conditions.

Once the starting values for the prognostic run are complete, monitoring can begin. If there are no initial values, for example stem positions are unknown, the missing data can be realistically complemented with the help of the stand structure generator (step 3).

Once the spatial model has been constructed (step 4) the silvicul-tural treatment program is specified in step 5. The competition index calculated for each tree through the three-dimensional model in step 6 is used, in step 7, to control individual tree development. Steps 4-7 are repeated until the entire prognostication period has been run through in 5-year steps.

To date, model research has had little success in substituting the yield tables for pure stands by an improved information system for pure and mixed stands. This can in no way be attributed to a deficit in methodological principles, data or technical equipment. Rather, the causes lie in the fact that new models are not properly adapted to practical requirements. The recent introduction of the growth model SILVA 2.2 for forest management use led to a range of operational requirements and outputs demanded from the management models that will be used in decision-making processes at stand and forest enterprise levels.

1 The natural management of forests is currently making great headway. In the long run only those growth models capable of simulating the growth of pure and mixed stands

of all age compositions and structural patterns will find approval.

2 Models need to be operable at stand and forest enterprise levels and able to simulate growth behaviour under different thinning regimes and different processes of artificial and natural regeneration.

3 Flexibility of the model is essential so as to permit simulation of growth reactions to site alterations and interference factors on a large regional scale.

4 Apart from tree and stand characteristics such as volume production, assortment yield, wood quality and financial yield should also include structural parameters determining the recreational and protective functions of forests as well as indicators showing the impact of hazards or ecological instability.

5 Forestry practice is interested, first and foremost, in calculating scenarios at stand and forest enterprise levels. This can only be achieved if input and output data of the model consider what information is available and which data are needed in forestry practice. Furthermore, achieving this goal also depends on whether the model forms part of a comprehensive forestry information system and, lastly, whether hardware specifications are acceptable in practice.

For decades forestry practice has been hoping for improved growth models to assist with research, planning, operations and control in forest management. The general acceptance of new models by practitioners calls for close cooperation between forest science and forest practice, from the design and development of the model to its actual introduction in forest management.

Likewise the only capital equipment required is a weaving needle, which is a very cheap item to buy. Similarly, wood is often used to make a variety of household implements and agricultural tools, or is carved to make products aimed at the tourist market. Once again, interviews with resource makers stress that the skills used to make these objects can be acquired in a very short time and the capital equipment required is minimal. Finally, where NTFP foods are processed to make commercial products, such as fruit wines, palm wines, dried meats, dried fish and dried insects, again the skills and capital equipment required are low: it is essentially a matter of allocating sufficient household labour to complete the task. As discussed earlier, due to the difficulties that rural households face in capital accumulation, because formal credit markets generally fail in rural areas and because education levels are low, rural households are usually endowed with unskilled labour but poorly endowed with capital. So another reason for the extensive use of NTFPs by rural households is that the factor inputs required to collect and process NTFPs (unskilled labour, little capital equipment) match closely the factor endowments of the classic rural household. (These required factor inputs, combined with open access to the NTFP resource, means that NTFP activities have very low entry barriers. In such circumstances we would therefore expect the returns to NTFP activities to be correspondingly low.)

Fourth, households also use NTFPs in response to the general riskiness of rural economic activities. The use of NTFPs, particularly what have been classified as 'minor forest products', during times of household stress is an observation common to much of the NTFP case study literature. NTFPs display a certain degree of non-covariance with respect to agricultural output. When crops fail due to drought or disease, or when shocks hit the household such as unemployment, death or disease, some NTFPs will still be available for the household to either consume or use to generate cash income to purchase its essential needs. Thus it is economically rational for risk-averse rural households to hold a portfolio of production opportunities that exploits this non-covariance. In this sense, commonly held forests and woodlands can be regarded as providing a set of back-stop resources that insure households against the failure of other, higher return production activities. Thus, even if the returns per hectare of woodland were systematically below that of agricultural land, rural villages would still retain common lands for this purpose.

Finally, the argument comes full circle. One of the reasons that rural households use NTFPs so extensively is precisely because they are open-access resources. There are a number of reasons for the survival of communally held resources, for example the insurance element noted above; however, a major reason for the survival of such resources is the very low income levels of rural households. Resource privatization is an expensive business: households taking charge of privatized resources usually need to spend time and money on creating exclusion (e.g. by building fences), on enforcement (e.g. employing guards to monitor incursions) and on punishing infractions (by going through the courts). Particularly where a resource has previously been held in common, privatization is often strongly contested and consequently these various costs can be expected to be high. Where households are generally poor, the private costs of resource privatization are likely to be far higher than the potential private gains from control of the resource. Thus the low incomes of rural areas underpin the existence of commonly held resources, which in turn underpins the extensive use of NTFPs by those very same low-income households (on the relationship between efficient property rights regimes and transactions costs, see Coase 1960; Bromley 1989; de Meza & Gould 1992).

## CAN RURAL DEVELOPMENT BE BASED AROUND NTFPS?

We have presented evidence that rural households' use of NTFPs is widespread and significant in terms of their overall livelihoods. We have also suggested that the cause of this extensive NTFP use by poor rural households is the match between the economic characteristics of NTFPs and the economic endowments and constraints of those same rural households. Can we therefore conclude from this that the commercialization of NTFPs is economically feasible and likely to bring substantial economic benefits to rural poor? Further, if commercialization were feasible, what would it be likely to do to the resource base that produces these NTFPs? In other words, what light does the analysis above shed on the possibility of using NTFPs to meet the twin goals of economic development and environmental conservation? I argue that there are substantial difficulties here, arising from the same economic characteristics of rural households and the causes of their NTFP use that were listed earlier. I look at a number of issues in turn and draw on the case study literature to illustrate where they have led to problems for various NTFPs.

## The problem of preferences

We saw earlier that one of the reasons for rural households' NTFP use was their low incomes and the still-partial monetization of many of their economic transactions. However, it is a truism that as individuals and households become more affluent, their demands for goods and services change. In particular, people choose to consume more highly preferred goods or goods of a higher quality. (There is abundant evidence for these demand shifts in the higher-income developed countries.) However, evidence from academic studies and from history suggests that NTFPs do not have high product quality or are not in general highly preferred goods. One way of showing this is to look at income elasticities of demand for NTFPs, as these elasticities reveal what the demand response of households will be to any increase in their incomes.

Unfortunately, demand elasticities for NTFPs are not easy to come by, due to the lack of attention paid to such goods by formal economic studies. Thus we have culled evidence from two sources. The first is the author's own study of households and environmental resources in Shindi, Zimbabwe and the second is from a recent literature review reported in Köhlin (1998) (Tables 15.5 & 15.6). Table 15.5 contains income elasticities for a range of NTFPs: wild fruits, wild vegetables, wild animals, firewood, NTFP-derived goods and thatching grass. In general, the income elasticities for these goods are low. In only one case, wild fish demands, is the recorded elasticity greater than 1 (i.e. a luxury good), although it is believed this has much to do with the supply conditions of wild fish in the study site. Otherwise most elasticities are closer to zero than to 1. For example, wild fruits collectively and mice have income elasticities of 0.3-0.4; woodland-derived wild vegetables have an elasticity of 0-0.2; firewood has an elasticity of 0.4; and NTFP-derived goods have income elasticities of 0.4-0.5. While these elasticities are positive, so that increases in income within the range of incomes spanned by the sample will increase overall NTFP demands, the budget shares of these goods will decline, to be replaced in importance by other more preferred goods. Table 15.5 gives one example of this, by comparing the income elasticities of woodland-derived and field-derived wild vegetables. The fact that the latter are much higher than the former reflects the higher preference households have for these foods, and also means that as their incomes rise rural households will switch to field-derived wild vegetables and away from their woodland-derived counterparts.

Further, the positive income elasticities in this table for NTFPs reflect the rather low income levels of Shindi households. Evidence from the regressions suggests that at higher income levels these elasticities would turn negative, so that at higher incomes these NTFPs would actually be inferior goods.

While this table draws on the wider literature on NTFPs, most of the quoted elasticities refer to firewood, as this has been the main quantitative focus of the literature. Three studies of firewood demands conclude that income elasticities turn negative as incomes rise (Amacher *et al.* 1993 for Nepal; Bagahawatte 1997 for Sri Lanka; Shyamsundar & Kramer 1996 for Madagascar). Further, the other elasticities cited are all either low and positive, or negative. This implies that as households get richer, the absolute quantities of NTFPs that they wish to consume will decrease until at some point NTFP demands will fall to zero. This evidence from formally estimated income elasticities is also backed up by classic observations of consumption shifts that occur as rural households become richer. For example, one immediate shift is from natural construction materials (earth bricks, poles, thatching grass) towards higher-quality purchased substitutes (breeze blocks, zinc roofing). Likewise, richer households choose to consume more purchased foods and beverages and less wild foods and locally produced alcohols. As the income elasticities suggested, richer households also prefer to switch away from firewood towards other sources of energy and lighting.

There are good economic reasons why NTFPs are regarded as inferior goods by richer households. Their production is variable, the quality of the goods cannot be controlled and other substitutes have been bred or produced with higher product quality in mind. Thus more affluent households shift to better substitutes when they can afford to do so. However, the key point is that the inferiority of NTFP goods creates a real problem for NTFP commercialization projects. Dependence on, and use of, NTFPs are linked to poverty and to market failure rather than to household choice: the current prevalence of NTFP use by rural households is a result of their low incomes rather than the attraction of NTFP products themselves. A dramatic illustration of this is given by Hegde *et al.* (1996), who argue that the very high rates of NTFP utilization of their sample households are caused by the fact that these households earn barely more than the Indian minimum wage, and present empirical evidence that as households get richer so their NTFP use declines. So where NTFPs are inferior, attempts to commercialize them will founder on a lack of long-term product demand.

## High transactions costs of trading

In our discussion of the stylized features of rural areas, we stressed the role that economic remoteness played in underpinning low rural incomes and in causing the failure of formal product markets. This is also a major cause of the low monetization of rural households' activities, since trading in formal markets outside the rural area is so often uneconomic. These transactions costs will affect NTFP commercialization projects just as much as they would affect any other project aimed at commercializing rural produce. The high unit costs of transportation imply that NTFPs which have a low unit value relative to their weight will have low if not negative margins in the marketplace.

This may be the case for a number of NTFPs. On account of the open-access nature of the resource rules surrounding forests and woodlands in developing countries and the fact that many are inferior goods, NTFP prices tend to be fairly low. Some NTFPs will have high weight-to-price ratios, especially those made of wood. For example, it is noticeable that despite the reasonable prices being attained for carvings on the major roads of Zimbabwe, production activity is contained within a reasonably narrow area, close enough to the road to make transporting the wood inputs an economic proposition. High trading costs have also been implicated in the lack of NTFP commercialization in Latin America (Richards 1993). Of course, the converse of this point is the effect that road building has on reducing trading costs and thereby stimulating the marketization of rural produce generally. The impact of the Belem-Brasilia and the Transama-zon highways on economic conditions in the Amazon basin has been widely noted. However, assuming that NTFP projects take the physical infrastructure as given, the impact of transactions costs on the profitability of NTFP commercialization will remain a serious constraint for remoter rural areas for a number of NTFPs.

## Storage problems

Another feature of rural areas we noted was the generally low levels of physical capital. This is true both at the household level and at the community level. Rural areas not only lack household-based productive capital but also the area-wide capital stock, such as power supplies and communications, on which the productivity of household-based enterprises depends. One key problem that this raises is a lack of appropriate storage facilities, such as an absence of refrigeration and refrigerated transport. Indeed, storage of goods is often a major difficulty for rural households. Even for longer-lasting foods such as grains, losses are regularly incurred thanks to mould, rotting and crop pests despite the careful construction of granaries and the use of locally available goods to line and protect these granaries.

Problems with storage could be a major constraint to the commercialization of NTFPs where these goods are perishable. A substantial fraction of the NTFPs examined under the 'Hidden Harvest' programmes are exactly these types of goods: fruits, fruit-based beverages, wild meats and even wild fish. All these would require reasonably capital-intensive storage if commercialization projects were to succeed. Appropriate storage provides two major economic functions, the absence of which would cause problems to emerge. The first function is that of product preservation. Given that the commercialization of NTFPs requires transportation of products to new markets, an absence of adequate storage would result in high losses of products due to spoilage (rotting, bruising in transit, excess fermentation of liquids, etc.). The other function that storage can play is that of interseasonal supply smoothing. Without it, products can only be supplied to the market during the season of their production, resulting in glutted markets, lower product prices and unsmoothed income streams. Thus seasonality and perishability lower the economic returns of NTFPs to rural communities, thereby making these activities less attractive.

## Production risk

As discussed earlier, rural households face considerable levels of risk, and one reason for the preservation of woodlands for NTFPs is as a risk-spreading insurance mechanism against severe shocks to the other economic activities of the household. However, in the medium to long term risk-averse rural households will wish not just to spread risk but to reduce it. If NTFP activities themselves still involve considerable production risks, then such activities will be less attractive to rural households than other, more certain income flows. This will be a consideration where the supply of NTFPs is dependent on climatic conditions, such as some of the foods mentioned earlier, or where NTFPs are prone to disease.

## Open access and the costs of privatization

A further concern derives from the fact that many forests and woodlands in developing countries are held under communal tenure, with usufruct rights to NTFPs very often being effectively open access. If NTFPs were to be commercialized, then one immediate concern would relate to the degree of exploitation of the resource. Raising the value of an NTFP will naturally increase supply pressures on the resource. Where NTFP use is nondestructive to the resource stock, this is not necessarily a problem. However, where NTFP use can diminish the resource stock, then the commercialization of the resource can have serious implications for objectives of environmental conservation. Indeed, there are a number of cases where increases in resource pressure consequent on increased demand for NTFP goods has had serious environmental consequences. One such case is the impact of basket-making on an indigenous tree in Botswana. The main source of leaves for baskets is *Hyphaene petersiana,* which stands up very well to intense harvesting pressure. However, the main source of dye for these baskets is the bark of *Berchemia discolor,* and harvesting the bark can harm the tree if done too frequently. The increase in demand for Botswana baskets has increased substantially the demand for dye inputs, and this in turn has led to the death of many *B. discolor* in basket-making areas (Cunningham & Terry 1995). Similarly, the high demand for *Warbergia salutaris* as a source of medicines in Zimbabwe has led to the almost complete extinction of this species within the country (Mukamuri & Kozanai 1999). There are numerous cases of this dynamic in the Amazonian rain forests, where palm heart production has resulted in the near elimination of *Euterpe edulis* and *Euterpe oleracea*; the cutting of palms for fruits in Peru has rapidly depleted the numbers of *Mauritia flexuosa, Mauritellia peruviana* and *Jessenia bataua*; and the felling of trees for rosewood oil production had led to the virtual extinction of *Aniba rosaeodora* and *Aniba ducke* in economically accessible parts of Brazil (Richards 1993). Commercialization of a destructive NTFP use when held under open access can thus have severe environmental consequences for the species in question (though not necessarily for the forest as a whole).

A longer-term concern relates to the question of who gains from NTFP commercialization. We noted earlier that the existence of communal tenure over woodlands is related to the high costs of exclusion, monitoring and enforcement relative to the benefits to be gained from private woodland control. In other words, the tenurial arrangements over woodlands are related to the ratio of costs and benefits faced by rural households and communities. If as a result of NTFP projects the per-hectare returns to woodland rise, then this ratio changes and economic incentives will arise to change tenurial arrangements. In particular, privatization may emerge as a cost-effective option. However, this is only likely to be cost-effective for households that are already more affluent than others, since only these will be able to bear the private costs of resource control. Two groups stand out here. The first is local élites. There is evidence that as open-access resources rise in value, these resources become co-opted by local élites. The classic study of this phenomenon is Ensminger's (1990) analysis of resource privatization in rural Kenya. In response to the lowering of trading costs via cheaper transportation costs, communal grazing lands were privatized with the consent of rural elders who themselves benefited from this process. However, this process of élite partitioning of commons resources in response to changing economic conditions has also been found in India (Jodha 1986) and Brazil (Hecht 1985). The second group is outsiders. Town-dwellers or traders/middlemen with superior technology and access to capital may gain control of either the NTFP resources or the NTFP goods' trade and thereby arrogate to themselves the economic returns to these goods. In either case, commercialization-induced changes in local tenure could not only stymie the expected socioeconomic gains from NTFP products to the generality of rural households, but also by triggering the privatization of woodlands and other common resources may leave the rural poor worse off by removing from them a vital consumption back-stop and a source of insurance goods.

## Incentives for domestication and technical substitution

We have discussed the many reasons above why NTFPs are currently often low-value goods, such as missing markets, the prevalence of local trading, low local demand and the open-access nature of NTFP resources. One implication of these low values is that neither rural households nor outside agencies currently have an incentive to invest in any sort of resource management or product improvement, since the costs of such actions would far outweigh the benefits. However, this would change if NTFPs were to become higher-value goods.

There are two processes to mention here: domestication and the development of technical substitutes.[9] It is globally true that when natural resources become valuable, there is pressure to intensify production and to domesticate the resource. Take as an example a wild fruit. While prices for the fruit remain low, it is unprofitable for households to do anything other than collect the fruit from the wild. However, as fruit prices rise, at some threshold value it becomes economic for the household (or some other agency) to incur the time and resource costs of establishing the resource under their own control, through planting and tending the relevant species.

Furthermore, domestication also allows resource managers to improve product quality through the development of a constant, uniform or standardized product, thereby maintaining or even increasing product prices. This dynamic is the general story of the development of agricultural products and there is no reason why NTFPs shouid be any different. Indeed, there are ample examples of cases where rising NTFP values have led to precisely this response. Perhaps the most famous is that of Brazilian rubber *(Hevea brasiliensis)* collection from natural forests. As rubber prices rose in the early part of the century, so it became economic to establish rubber plantations in South-East Asia; more recently, rubber plantations have been established in Brazil as well and these now supply 60% of Brazil's market (Browder 1992). Another classic NTFP affected in this way is brazil nuts. In response to rising nut prices, several thousand hectares of *Berthol-letia excelsa* plantations have now been established within Brazil itself. So neither rubber nor brazil nuts provide a solid foundation for NTFP commercialization in the Amazon in the long run. Indeed, it is difficult to point to a wild resource which, once valuable, has remained undomesticated (an extensive list of other Amazonian NTFPs that have been domesticated is given by Homma 1992). Beyond a certain point, then, the commercialization of NTFPs will almost certainly lead to the product being removed from the forest and being placed on the farm instead.

Additionally, if NTFPs become high-value goods, there is an economic incentive to invest in the search for cheaper synthetic substitutes, if this is technically feasible. Once such substitutes are developed, the market for NTFPs often collapses - the 'bust' following the 'boom'. Richards (1993) discusses many such examples of substitute-induced market collapses in commercialized Amazonian NTFPs. At various times these have included dyewood *(Caesalpinia echi-nata)*, vegetable ivory *(Phytelephus macrocarpa)*, balata *(Manilkara bidentada)*, leche caspi *(Couma macrocarpa)*, barbasco *(Lonchocarpus* spp.), babaçu oil *(Orbignya phalerata)*, cumaru nuts *(Dipteryx odorata)*, sorva latex *(Couma utilis* and *C. rigida)* and of course rubber. Such technical substitutions are problems for the entire goal of NTFP-based economic development, since they raise the spectre of entire markets disappearing just as rural households have come to depend on the NTFP resource for a high proportion of their livelihoods.

## CONCLUSIONS

We should not therefore underestimate the problems associated with the use of NTFPs for the economic development of rural areas. Although rural households currently use NTFPs quite extensively, this is a consequence of the low incomes, capital scarcity, missing markets, riskiness in production and tenure conditions that characterize rural areas. It is these that make it economically rational for households currently to use NTFPs, as these are low value, depend on unskilled labour for their collection and processing, and offset to a degree the riskiness of other production activities.

However, this is no guarantee of success when it comes to the commercialization of NTFPs: (i) there is little evidence that NTFP dependence in the long run is desired at all by rural households; (ii) there are economic problems associated with the costs of both trading and storing NTFPs; (iii) NTFPs still have levels of production risk that may make them unattractive; (iv) the commercialization of NTFPs is likely to lead to the elimination of some species; (v) NTFP development may also undermine the communal tenure system, thereby removing the wider benefits of NTFP use from poorer households; and (vi) NTFP projects aimed at raising product prices will have to face the twin pressures of domestication and technical substitution that have affected almost all other wild products. In the face of these problems, NTFP extraction-based rural economies are likely to be highly unstable, thereby undermining the goal of improved rural household welfare.

# Bibliography


Aukland, L.: (1997). *The Effect of Selective Logging on Amphibian Diversity of Budongo Forest*, Uganda. B. A., Oxford University.

Bahati, J.: (1995). *Impact of Arboricidal Treatments on the Natural Regeneration of Species Composition in Budongo Forest*, Uganda. M. Sc., Makerere University.

Barrows, M.: (1996). *A Survey of the Intestinal Parasites of the Primates in Budongo Forest*, Uganda. Glasgow University.

Bethell, E.: (1998). *Vigilance in Foraging Chimpanzees.* University College, London University.

Fairgrieve, C.: (1995). *The Comparative Ecology of Blue Monkeys (Cercopithecus mitis stuhlmanni) in Logged and Unlogged Forest, Budongo Forest Reserve*, Uganda, Edinburgh University.

Johnson, K.: (1993). *Local Uses of Budongo's Forest Products.* M. Sc., Oxford University.

Johnson, K.: (1996). *Hunting in the Budongo Forest*, Uganda. SWARA.

Miller, P. L.: (1993). *Some dragonflies of the Budongo Forest*, Western Uganda. Opusc.

Mork, K. & Solvang, R.: (1996). *Edge Effects on an Understorey Bird Community in Budongo Forest Reserve*, Uganda, Agricultural University of Norway.

Reynolds, V.: (1965). *Budongo: A Forest and its Chimpanzees*, London, Methuen.

Reynolds, V. & Reynolds, F.: (1965). *Chimpanzees of the Budongo Forest*, New York, Rinehart and Winston.

Shiel, D.: (1997). *The Ecology of Long-term Change in a Ugandan Rainforest*. D. Phil., Oxford University.

# Index



□□□